Smart Nanocomposite's *letters* 2018

Editor in Chief: Kirill L. Levine   St. Petersburg Mining University

SMART NANOCOMPOSITE'S *letters*

Volume 1, 2018

TABLE OF CONTENTS

SMART NANOCOMPOSITE'S *letters*

Periodic Digest "**Smart Nanocomposite's** *letters* " continues edition of magazine "Smart Nanocomposites" issuing by Nova Science Publishers (USA, NY) since 2010.

Editor team of **Smart Nanocomposite's** *letters* are grateful to Nova Science Publishers for hosting journal from Volume 1, issue 1 to Volume 8, issue 2. At the same time we are honored to announce **Smart Nanocomposite's** *letters* as a new periodic Digest starting from Volume 1.

About

SMART NANOCOMPOSITE'S *letters*

This Digest presents new studies in the fast growing area of smart materials, in particular, composite nanostructured materials. It focuses on the physics and physical chemistry of surfaces, interfaces, thin films and coatings, nanoparticles and other nanostructures, as well as on their new and smart applications. Original approaches in fabrication and applications of nanostructured materials will get special attention. Nanostructured ceramics, alloys, various nanocarbon forms (nanotubes, fullerenes, graphene) and their composites used in sensors (including single molecule sensing) and actuators, artificial metabolism, drug delivery, selective membranes, fuel cells, energy storage, and photovoltaics are just a few examples of new classes of materials and applications that are within the scope of the Digest. It features the results of interdisciplinary research from universities, national labs, and privately owned companies.

The Digest is peer-reviewed with the highest standards and quality of publications. The purpose of this Digest is to bring the most up-to-date advances in nanotechnology together, and to give research groups the opportunity to compare their results with other groups' data. To achieve this, the Digest focuses mostly on practical applications of nanodevices, and on proof of the concept publications. Areas of interest include (but not are limited to): sensors, smart membranes, smart coatings for corrosion protection, aspects of significance to nanorobots: power supplies, nanorobot manipulating devices, and microchips for artificial intelligence. The Digest also deals with safety issues: safety of nanotechnology to the environment, controlling the nanodevices, and other aspects.

Smart Nanocomposite's letters
is published in two issues per year by

Science Impact
SC, U.S.A.

E-mail: science_impact@hotmail.com
ISBN: 978-1717149497

Subscription Rate per Issue: $90

Editor in Chief: Editor in Chief: Kirill L. Levine, St. Petersburg Mining University, St. Petersburg, Russia

EDITOR-IN-CHIEF

Dr. Kirill Levine

General and Technical Physics

St. Petersburg Mining University, Russia

COORDINATING EDITORS

Dr. Stanislav Moshkalev

Center for Semiconductor Components CCS

University of Campanos, Brasil

Dr. Dennis E. Tallman

North Dakota State University

Department of Chemistry and Biochemistry, Fargo, ND, USA

Professor Andrey G. Syrkov

General and Technical Physics

St. Petersburg Mining University, Russia

Professor Yury A. Gorokhovatsky

Department of General and Experimental Physics

Herzen University, St. Petersburg, Russia

Professor Alexandr S. Mustafaev

General and Technical Physics

St Petersburg Mining University, Russia

EDITORIAL BOARD MEMBERS

Dr. Inamuddin

Advanced Functional Materials Laboratory

Department of Applied Chemistry

Faculty of Engineering and Technology

Aligarh Muslim University, Aligarh- 202 002, India

Dr. Jude O. Iroh

Chemical and Materials Engineering

University of Cincinnati, USA

Dr. Mihaela Manea

Laboratory Engineer

The Mud Lab for Central Europe of M-I Swaco, Romania

Professor Nikolay S. Pshchelko

Academy of communications named after S.M. Budenny, St. Petersburg, Russia

Dr. Ricardo Santos

Faculdade de Engenharia da Universidade do Porto, Portugal

Prof. Dale W. Schaefer

Department of Biomedical, Chemical and Environmental Engineering

University of Cincinnati

Cincinnati, Ohio, USA

Features of metrological support of the implementation of nanotechnology taking into account nonlinear effects

A.G. Syrkov, A.A. Vinogradova, K.L. Levine*

*syrkovandrey@mail.ru

St. Petersburg Mining University

ABSTRACT

Discussed the historical tradition of development of devices and control methods for nano-objects and the modern features of metrological support of the implementation of nanotechnology.

Keywords: nanotechnology, research methods, additives

INTRODUCTION

In recent years the leading manufacturers of scientific equipment was developed, high-precision instruments for the measurement and integrated interpretation of micro - and nanoscale mechanical and tribological properties of nanostructured materials and coatings based on the methods of measuring the indentation (penetration testing), scratching (adhesion test) and friction (tribological tests).

DISCUSSION

For metrological measurements correct and commercialization of products of nanotechnology require a calibrated means of measurement, including standard samples intended for use in the system of ensuring the uniformity of measurements for verification, calibration, control of metrological characteristics of measuring instruments when conducting their tests, metrological certification of measurement procedures, monitoring the stability of the measurement results [1, 2, 3].

To date, only probe techniques using different nanoscopes, are unable to solve the problem of adequate control of surface-nanostructured metals and tribological materials based on them taking into account nonlinear effects. According to the author of the works [4, 5] to solve this problem, a promising approach is based on finding relationships between macroscopic properties, which are sensitive to changes in the composition of the surface layers of the substance.

For oil I-20 with additives dispersed metals containing on the surface of low-dimensional forms of ammonium compounds, the most sensitive proved to be an integral indicator of friction D measured in the range of acoustic frequencies, and its relationship with adsorption-chemical characteristics of dispersed additives [4-6].

In the Mining University this is more than a century long tradition of development of devices and control methods for nano-structured objects. It comes from P.P. Veimarn, its Professor and graduate. Some recognized experts in nanotechnology consider Veimarn as the founder of the science of nanotechnology, which is the first concept formulated basic principles of nanotechnology approach [7-9]. In the early twentieth century Weimarn actively used optical microscopy Sismondi and studied electrical conductivity of highly dispersed metals (1912) [10].

One of the main problems related to metrological assurance of nanotechnology is the problem of test object [2]. In the measurement of characteristics of nanoscale objects it is very difficult to observe the basic principle of Metrology: measuring instrument must have an accuracy high enough to provide valid measurement. The most difficult task is to find etalon (the transfer unit size) in the nanometer range. Practical implementation of finding etalon depend on the respective reference measurement [8, 9, 11]. According to [11], the most "prepared" method for solving the problem of control of the sizes of elements large-scale integrated circuits in microelectronics is the method of scanning electron microscopy; atomic force microscopy also is promising to solve the considered task.

CONCLUSION

Further developing approach described by Weimarn , the problems arising in the measurement of nanometric objects was being solved. It made possible measuring D (the integral indicator of friction) in the range of acoustic frequencies, which is indicative to structure of monomolecular surface layer of the additive [12, 13].

Therefore, it is advised to use friction method in the range of acoustic frequencies, as it was made to achieve full control of tribological parameters and make diagnostics of the tribological system with surface-modified additives [14].

REFERENCES

[1] Blank, V. D. Gogolinsky K. V., Pyatov, A. L. Legislative and applied Metrology. 1 (107), (2010) 27-28.

[2] Kovalchuk M. V., Todua P. A. Technosphere, (2009) 136.

[3] Gogolinsky K. V. Proceedings of the Mining Institute, (2015) 264.

[4] Syrkov A. G. Proceedings of the Mining Institute, 216, 2015. 122-130.

[5] Syrkov A. G. Publishing house of Polytechnic University, (2016) 194.

[6] Nazarova E. A., Syrkov A. G., Brichkin V. N., Advanced Materials Research, 1040, (2014) 103–106.

[7] Zhabrev V. A., Publishing House "Elmor", (2012) 328.

[8] Margolin V. I. Publisher "LAN", (2012) 27.

[9] Syrkov A. G. Condensed matter and the interface, 19 (1), (2017) 140-147.

[10] Veimarn, P. P. Econ. Tipo-Litor., 1912. 15.

[11] Novikov Y. A., Rakov A.V. Measuring equipment, 1, (1999) 14-18.

[12] Vinogradova A. A. Syrkov A. G., Simakov A. S. Condensed matter and interphases, 15 (2), (2013) 203-205.

[13] Vinogradova A. A., Boikov A. V., Syrkov A. G. Competence, 7, (2016) 12-16.

[14] Boikov A. V., Vinogradova A. A., Syrkov A. G. Competence, 9-10, (2016) 203-205.

Synthesis, Identification and Physical-Chemical Properties of bis-Adduct of Fullerene C60 with Histidine

Vlada V.Petrenko[a], Nikolay A.Charykov[a,b], Konstantin N.Semenov[a,c], Viktor A.Keskinov[a], Aleksey V.Kurilenko[a], Irina V.Semenyuk,[a] Nikolay M.Safiannikov[b]

keskinov@mail.ru

[a]Saint-Petersburg State Technological Institute (Technical University), Russia

[b]Saint-Petersburg State Electro-technical University (LETI), Russia

[c]Saint-Petersburg State University, Russia

ABSTRACT

The heterogeneous − non-catalytic method of the synthesis and purification (by water − methanol re-crystallization) of bis-adduct of fullerene C_{60} with amino-acid histidine is described. Identification of bis-adduct was provided by the following methods of physical-chemical analysis: C-H-N-O Element analysis, High performance liquid chromatography, IR-spectroscopy, Electronic spectroscopy. Volume properties of bis-adduct of fullerene C_{60} with histidine water solutions - $C_{60}(C_6N_3H_8O_2)_2(H)_2$ − H_2O at 298 K (concentration dependencies of density, average and partial molar volumes of both components) were determined with the help of quartz pycnometers. Concentration dependencies of refraction properties (refraction indexes, specific and molar refractions of the solutions and bis-adducts) at 298 K were determined with the help of refractometers. Using method of isothermal saturation in ampules, poly-thermal solubility of $C_{60}(C_6N_3H_8O_2)_2(H)_2$ in water solutions was determined in the temperature range 293 − 353 K. By the method of visible light scattering, concentration dependencies of linear dimensions of nano-cluster associates and electro-kinetic ζ − potentials were determined at 298 K. By the method of cryo-metry with the help of Beckmann thermometer concentration dependence of the decrease of liquidus temperatures was determined at the temperatures nearby 273 K. On the base of last data excess function of both components - $C_{60}(C_6N_3H_8O_2)_2(H)_2$ andH_2O (activities and activity coefficients) were calculated, using novel semi-empirical Virial Decomposition Asymmetric Model (VD-AS). Also with the help of VD-AS diffusional instability concentration region was calculated.

15

Keywords: bis-adduct, fullerene C_{60}, histidine, synthesis, heterogeneous – non-catalytic method, identification, physical-chemical analysis, water solutions, physical-chemical properties.

INTRODUCTION

Water soluble derivatives of fullerenes is a perspective class of compounds due to possibilities of their application in various fields of science and technology especially in biology and medicine due to well definedmembranotrophic, cytoprotective, radioprotective, antioxidant, antimicrobial, antiviral and transport properties [1–3].This chapter is devoted to the development of the cycle of works, devoted to the synthesis, identification and the investigation pf physical-chemical properties of the adducts of light fullerenes (C_{60} or C_{70}) and amino-acids: alanine, lysine, histidine, arginine, glycine, proline etc. Stromet prepared fullerene aminoacids by dipolar addition to C_{60} of either the Boc- or Fmoc-Na-protected azido amino acids derived from phenylalanine and lysine [4]. It was determined that prepared aminoacids are a mixture of 5,6-open (major product) and 6,6-closed (minor product) derivatives. Kotelnikova et al. synthesized the water-soluble C60 derivatives with DL-alanine and investigated the influence of the obtained compounds on structure and permeability of the lipid bilayer of phosphatidylcholine liposomes [5].Hu et al. synthesized the C60 derivative with β-alanine, cystine and arginine [6].Kumar et al. synthesized and studied DNA cleavage efficiency upon visible light (in the presence of nicotinamide adenine dinucleotide) ofthe C60-lysine derivative [7].Jiang et al. synthesized novel water-soluble C60-glycine derivative [8]. The results of cytotoxicity assay of cancer cell lines showed thatC60-glycine derivative in a dose-dependent manner increases cell death. Authors of [9,10] synthesized hybrid structures based on fullereneC60 with attached proline amino acid (methyl ether of N-[(β-alanylhistidyl-ethyl) fullerenyl] proline, methyl ether of N-[(nitroxyethyl)fullerenyl] proline, methyl ether of N-[(2',3'-dinitroxy-propyl)fullerenyl] proline, methyl ether of N-[mono-hydro-fullerenyl] proline, carnosine). It was determined that all studied compounds possess anantioxidant activity and inhibited glutamate induced Ca^{2+} uptake into synapto somes of the rat brain cortex. Some articles were fulfilled with the participation of the authors of the presented chapter [11-17].

MATERIALS AND METHODS

The heterogeneous – non-catalytic method of the synthesis of bis-adduct of fullerene C_{60} with amino-acid histidine -$C_{60}(C_6H_8N_3O_2)_2(H_2)$ or is described. Histidine (mass m=3.2 g), base NaOH (m=15.8 g) were dissolved in 54 cm^3 of H_2O; 130 cm^3 of the solution of fullerene C_{60} in o-xylene with the concentration 7.62 $g/dm^3$was prepared by direct dissolution. Then ethanol (C_2H_5OH) (m=267 g) was added to the both "o-xylene" and "water" phases and the heterogeneous mixture was stirred for 7 days. Then phases were divided, "water" phase was evaporated and bis-adduct $C_{60}(C_6H_8N_3O_2)_2(H_2)$ was precipitated by methanol (CH_3OH). Purification of bis-adduct was provided by triple recrystallization from water solutions by methanol. Additionally bis-adduct $C_{60}(C_6H_8N_3O_2)_2(H_2)$ was washed from the impurities of sodium-contain forms by HCl acid in Soxhlet apparatus. Yield of bis-adduct was 76 % from the theoretical one.

Identification.

Identification of bis-adduct $C_{60}(C_6H_8N_3O_2)_2(H_2)$ was provided by the following methods of physical-chemical analysis:

- C-H-N-O Element analysis (CHNS/O Analyzer 2400 Series II) proved formula of crystal hydrate $C_{60}(C_6N_3H_8O_2)_2(H)_2*24H_2O$;

- High performance liquid chromatography (HPLC PerkinElmer) confirmed purity of adduct 96.5 mass %;

- IR-spectroscopy (IRTracer-100, Shimadzu) in wavenumbers: $v = 400 - 4000$ cm^{-1} demonstrates absorption characteristic peaksfor: valent O-H, C-H, CH_2, C=O, deformation O-H, N-H oscillations and long-wave characteristic oscillations of C_{60} fullerene core (5009 – 1000 cm^{-1});

- Electronic spectroscopy (UV-1800, Shimadzu), wavelength: $\lambda = 190$-1100 nm demonstrates no effects and monotonically increasing with wavelength decreasing light absorption.

EXPERIMENTAL SECTION

Volume properties of bis-adduct of fullerene C_{60} with histidine water solutions - $C_{60}(C_6N_3H_8O_2)_2(H)_2$ – H_2O at 298 K (concentration dependencies of density, average and partial molar volumes of both components) were determined with the help of quartz pycnometers. Concentration dependencies of refraction properties (refraction indexes, specific and molar refractions of the solutions and bis-adducts) at 298 K were determined with the help of refractometers. Using method of isothermal saturation in ampules, poly-thermal solubility of $C_{60}(C_6N_3H_8O_2)_2(H)_2$ in water solutions was determined in the temperature range 293 – 353 K. By the method of visible light scattering, concentration dependencies of linear dimensions of nano-cluster associates and electro-kinetic ζ – potentials were determined at 298 K. By the method of cryo-metry with the help of Beckmann thermometer concentration dependence of the decrease of liquidus temperatures was determined at the temperatures nearby 273 K. On the base of last data excess function of both components - $C_{60}(C_6N_3H_8O_2)_2(H)_2$ and H_2O (activities and activity coefficients) were calculated, using novel semi-empirical Virial Decomposition Asymmetric Model (VD-AS). Also with the help of VD-AS diffusional instability concentration region was calculated.

ACKNOWLEDGEMENTS

This work was supported by Russian Found of Fundamental Investigations – RFFI (Projects no. 16-08-01206, 18-08-00143 A),by the Grant of the President of Russian Federation for supporting young scientists MK- 4657.2015.3. Research was performed with using the equipment of the Resource Centers 'GeoModel', Center for Chemical Analysis and Materials Research and Center for Thermogravimetric and Calorimetric Research of Research park of St. Petersburg State University.

REFERENCES

[1] F. Cataldo, T. Da Ros, Carbon Materials: Chemistry and Physics: Medical Chemistry and Pharmacological Potential of Fullerenes and Carbon Nanotube, Springer, 2008.

[2] L.N. Sidorov, M.A. Yurovskaya, Fullerenes, Ekzamen, Moscow, 2005.

[3] L.B. Piotrovskii, O.I. Kiselev, Fullerenes in Biology, 2006 (Rostok, Saint-Petersburg).

[4] T.A. Strom, A.R. Barron, Chem. Commun. 46 (2010) 4764–4766.

[5] R.A. Kotelnikova, A.I. Kotelnikov, G.N. Bogdanov, V.S. Romanova, E.F. Kuleshova, Z.N.Parnes, M.E. Volpin, FEBS Lett. 389 (1996) 111–114.

[6] Z. Hu,W. Guan,W.Wang, L. Huang, X. Tang, H. Xu, Z. Zhu, X. Xie, H. Xing, Carbon 46. (2008) 99–109.

[7] A. Kumar, M.V. Rao, S.K. Menon, Tetrahedron Lett. 50 (2009) 6526–6530.

[8] G. Jiang, F. Yin, J. Duan, G. Li, J. Mater. Sci. Mater. Med. 26 (2015) 1–7.

[9] V.V. Grigoriev, L.N. Petrova, T.A. Ivanova, R.A. Kotelnikova, G.N. Bogdanov, D.A., Poletayeva, I.I. Faingold, D.V. Mishchenko, V.S. Romanova, A.I. Kotel'nikov, S.O.Bachurin, Biol. Bull. 38 (2011) 125–131.

[10] L.V. Tatyanenko, O.V. Dobrokhotova, R.A. Kotelnikova, D.A. Poletayeva, D.V., Mishchenko, I.Y. Pikhteleva, G.N. Bogdanov, V.S. Romanova, A.I. Kotelnikov,Pharm. Chem. J. 45 (2011) 329–332.

[11] M.Y. Matuzenko, D.P. Tyurin, O.S. Manyakina, K.N. Semenov, N.A. Charykov, K.V.Ivanova, V.A. Keskinov, Nanosystems: Phys. Chem. Math. 6 (2015) 704–714.

[12] M.Y. Matuzenko, A.A. Shestopalova, K.N. Semenov, N.A. Charykov, V.A. Keskinov,Nanosystems: Phys. Chem. Math. 6 (2015) 715–725.

[13] A.A. Shestopalova, K.N. Semenov, N.A. Charykov, V.N. Postnov, N.M. Ivanova, V.V.Sharoyko, V.A. Keskinov, D.G. Letenko, V.A. Nikitin, V.V. Klepikov, I.V. Murin, J.Mol.

Liq. 211 (2015) 301–307.

[14] O.S. Manyakina, K.N. Semenov, N.A. Charykov, N.M. Ivanova, V.A. Keskinov, V.V., Sharoyko, D.G. Letenko, V.A. Nikitin, V.V. Klepikov, I.V. Murin, J. Mol. Liq. 211. (2015) 487–493.

[15] K.N. Semenov, N.A. Charykov, I.V. Murin, Y.V. Pukharenko, J. Mol. Liq. 201 (2015). 50–58.

[16] K.N. Semenov, N.A. Charykov, V.N. Keskinov, J. Chem. Eng. Data 56 (2011) 230–239.

[17] K.N. Semenov, N.A. Charykov, I.V. Murin, Y.V. Pukharenko, J. Mol. Liq. 201 (2015). 1–8.

Tris-adduct of light fullerene C_{70} with indispensable amino-acid lysine

Irina V.Semenyuk[a], Konstantin N.Semenov[a,c], Nikolay A.Charykov[a,b],

Viktor A.Keskinov[a], Aleksey V.Kurilenko[a], Vlada V.Petrenko[a],

Nikolay M.Safiannikov[b]

[a]Saint-Petersburg State Technological Institute (Technical University), Russia,

[b]Saint-Petersburg State Electro-technical University (LETI), Russia,

[c]Saint-Petersburg State University, Russia.

ABSTRACT

The heterogeneous – non-catalytic method of the synthesis and purification (by water – methanol re-crystallization) of tris-adduct of fullerene C_{70} with amino-acid lysine is described. Identification of tris-adduct was provided by the following methods of physical-chemical analysis: C-H-N-O Element analysis, High performance liquid chromatography, IR-spectroscopy, Electronic spectroscopy, Complex thermal analysis. Volume properties of tris-adduct of fullerene C_{70} with lysine water solutions - $C_{70}(C_6H_{13}N_2O_2)_3(H_3)$ at 298 K (concentration dependencies of density, average and partial molar volumes of both components) were determined with the help of quartz pycnometers. Concentration dependencies of refraction properties (refraction indexes, specific and molar refractions of the solutions and bis-adducts) at 298 K were determined with the help of refractometers. By the method of visible light scattering, concentration dependencies of linear dimensions of nano-cluster associates and electro-kinetic ζ – potentials were determined at 298 K. By the method of cryo-metry with the help of Beckmann thermometer concentration dependence of the decrease of liquidus temperatures was determined at the temperatures nearby 273 K. On the base of last data excess function of both components - $C_{70}(C_6H_{13}N_2O_2)_3(H_3)$ and H_2O (activities and activity coefficients) were calculated, using novel semi-empirical Virial Decomposition Asymmetric Model (VD-AS). Concentration dependencies of density, dynamic and kinematic viscosity at 293, 313 and 333 K was determined by the Stokes method of "falling gold ball" in water solutions of tris-adduct at 298 K.

Keywords: tris-adduct, fullerene C_{70}, lysine, synthesis, heterogeneous – non-catalytic method, identification, physical-chemical analysis, water solutions, physical-chemical properties.

INTRODUCTION

It is well-known, that light fullerenes (C_{60} and C_{70}) are practically are insoluble in water or water solutions or fluids, such as: physiological solutions, blood, lumph, CSF, gastric juice, food and wine products etc. This fact considerably limited application of light fullerenes and their derivatives in biology, medicine, pharmacology, cosmetics, food industry etc. despite of the fact that the last ones posses a number of very useful and rare properties, in particular: photo-protective, cytoprotective, radioprotective, antioxidant, bactericide, antiviral and transport. A wide range of articles were devoted to the synthesis, identification, physical-chemical, biological and medicine investigation in multicomponent systems with the adducts of light fullerenes with interchangeable and indispensable amino-acids (see, for example articles [1-14]).

MATERIALS AND METHODS

The heterogeneous – non-catalytic method of the synthesis of tris-adduct of fullerene C_{70} with amino-acid lysine - $C_{70}(C_6H_{13}N_2O_2)_3(H)_3$is described. Histidine (mass m=4.5 g), base NaOH (m=15.8 g) were dissolved in 54 cm^3 of H_2O; 130 cm^3 of the solution of fullerene C_{70} in o-xylene with the concentration 8.9 $g/dm^3$was prepared by direct dissolution. Then ethanol (C_2H_5OH) (m=250 g) was added to the both "o-xylene" and "water" phases and the heterogeneous mixture was stirred for 7 days. Then phases were divided, "water" phase was evaporated and tris-adduct $C_{70}(C_6H_{13}N_2O_2)_3(H)_3$was precipitated by methanol (CH_3OH). Purification of tris-adduct was provided by triple recrystallization from water solutions by methanol. Additionally tris-adduct $C_{70}(C_6H_{13}N_2O_2)_3(H)_3$was washed from the impurities of sodium-contain forms by HCl acid in Soxhlet apparatus. Yield of tris-adduct was 79 % from the theoretical one.

EXPERIMENTAL SECTION

Identification.

Identification of tris-adduct $C_{70}(C_6H_{13}N_2O_2)_3(H)_3$ was provided by the following methods of physical-chemical analysis:

- C-H-N-O Element analysis (CHNS/O Analyzer 2400 Series II) proved formula of crystal hydrate $C_{70}(C_6H_{13}N_2O_2)_3(H)_3*18H_2O$;

- High performance liquid chromatography (HPLC PerkinElmer) confirmed purity of tris-adduct 98.6 mass %;

- IR-spectroscopy (IRTracer-100, Shimadzu) in wavenumbers: v = 400 – 4000 cm^{-1} demonstrates absorption characteristic peaks for: valent O-H, C-H, CH_2, C=O, deformation O-H, N-H oscillations and long-wave characteristic oscillations of C_{60} fullerene core (5009 – 1000 cm^{-1});

- Electronic spectroscopy (UV-1800, Shimadzu), wavelength: λ = 190-1100 nm demonstrates no effects and monotonically increasing with wavelength decreasing light absorption;

- Complex thermal analysis shows stepwise external dehydration of crystal hydrates at 90 -210°C, stepwise decarboxylation, internal dehydration, denitrogenation at 300-810°C, C_{70} fullerene core oxidation at the temperatures higher than 900°C.

Physical-chemical properties.

Volume properties of tris-adduct of fullerene C_{70} with lysine water solutions - $C_{70}(C_6H_{13}N_2O_2)_3(H_3)$ at 298 K (concentration dependencies of density, average and partial molar volumes of both components) were determined with the help of quartz pycnometers. Concentration dependencies of refraction properties (refraction indexes, specific and molar refractions of the solutions and bis-adducts) at 298 K were determined with the help of refractometers. By the method of visible light scattering, concentration dependencies of linear dimensions of nano-cluster associates and electro-kinetic ζ – potentials were determined at 298 K. By the method of cryo-metry with the help of Beckmann thermometer concentration dependence of the decrease of liquidus temperatures was determined at the temperatures nearby 273 K. On the base of last data excess function of both components - $C_{70}(C_6H_{13}N_2O_2)_3(H_3)$ and H_2O (activities and activity coefficients) were calculated, using novel semi-empirical Virial Decomposition Asymmetric Model (VD-AS). Also with the help of VD-AS diffusional instability concentration region was calculated. Concentration dependencies of density, dynamic and kinematic viscosity at 293, 313 and 333 K was determined by the Stokes method of "falling gold ball" in water solutions of bis-adduct at 298 K. Concentration dependencies of hydrogen indicator (with the help of glass electrode), specific electro-conductivity (by RLC meter) were determined in tris-adduct water solutions at 298 K. On the base of both last data seeming dissociation degrees and concentration dissociation constants were calculated. Potential-metric titration of - $C_{70}(C_6H_{13}N_2O_2)_3(H_3)$ water solutions by acid (HCl) and base (NaOH) were provided.

ACKNOWLEDGEMENTS

This work was supported by Russian Found of Fundamental Investigations – RFFI (Projects no. 16-08-01206, 18-08-00143 A),by the Grant of the President of Russian Federation for supporting young scientists MK- 4657.2015.3. Research was performed with using the equipment of the Resource Centers 'GeoModel', Center for Chemical Analysis and Materials Research and Center for Thermogravimetric and Calorimetric Research of Research park of St. Petersburg State University.

REFERENCES

[1] Liang Bing GAN, Chu Ping LUO. Chinese Chemical letters. 1994.Vol.5, No.4, p 275-278.

[2] L.N. Sidorov, M.A. Yurovskaya, Fullerenes, Ekzamen, Moscow, 2005.

[3] A. Kumar, M.V. Rao, S.K. Menon, Tetrahedron Lett. 50 (2009) 6526–6530.

[4] R.A. Kotelnikova, A.I. Kotelnikov, G.N. Bogdanov, V.S. Romanova, E.F. Kuleshova, Z.N.Parnes, M.E. Volpin, FEBS Lett. 389 (1996) 111–114.

[5] Z. Hu,W. Guan,W.Wang, L. Huang, X. Tang, H. Xu, Z. Zhu, X. Xie, H. Xing, Carbon 46. (2008) 99–109.

[6] L.B. Piotrovskii, O.I. Kiselev, Fullerenes in Biology, 2006 (Rostok, Saint-Petersburg).

[7] G. Jiang, F. Yin, J. Duan, G. Li, J. Mater. Sci. Mater. Med. 26 (2015) 1–7.

[8] O.S. Manyakina, K.N. Semenov, N.A. Charykov, N.M. Ivanova, V.A. Keskinov, V.V., Sharoyko, D.G., Letenko, V.A. Nikitin, V.V. Klepikov, I.V. Murin, J. Mol. Liq. 211. (2015) 487–493.

[9] K.N. Semenov, N.A. Charykov, I.V. Murin, Y.V. Pukharenko, J. Mol. Liq. 201 (2015). 1–8.

[10] K.N. Semenov, N.A. Charykov, V.N. Keskinov, J. Chem. Eng. Data. 56 (2011) 230–239.

[11] A.A. Shestopalova, K.N. Semenov, N.A. Charykov, V.N. Postnov, N.M. Ivanova, V.V.Sharoyko, V.A. Keskinov, D.G. Letenko, V.A. Nikitin, V.V. Klepikov, I.V. Murin, J.Mol. Liq. 211 (2015) 301–307.

[12] K.N. Semenov, N.A. Charykov, I.V. Murin, Y.V. Pukharenko, J. Mol. Liq. 201. (2015). 50–58.

[13] Semenov K.N., Iurev G.O., Ivanova N.M., Postnov V.N., Sharoyko V.V., Prikhodko I.V., Murin I.V., Charykov N.A., Keskinov V.A., LetenkoD.G., Kulenova N.A.J.Mol.Liq.225. (2017). 767-777.

[14] Matuzenko M.Yu., Shestopalova A.A., Semenov K.N., Charykov N.A., Keskinov V.A.
Nanosys.: Ph., Ch., Math. 6. (2015). № 5. 715-725.

Smart Nanocomposite's *letters*
Volume 1   © Science Impact

Dielectric relaxation and molecular mobility in thermoplastic aromatic polyimides with different structures

*Natalia A. Nikonorova[1], Rene A. Castro[2], Dao T. Hong[2]**

and Alexei A. Kononov[2]

[1]Institute of Macromolecular Compounds of Russian Academy of Sciences, St. Petersburg, 199004 Russia

[2]Herzen State Pedagogical University of Russia, St. Petersburg, 191186 Russia

*honghip2012@gmail.com

ABSTRACT

The results of study of the dielectric relaxation and molecular mobility in five thermoplastic aromatic polyimides with different structures are reported. Three regions of dielectric loss due to relaxation of dipole polarization: the γ and β (in the glassy state) and the α (in the highly elastic state) were found. The key dielectric relaxation parameters are determined for different temperature ranges.

Keywords: thermoplastic aromatic polyimides, dielectric relaxation, molecular mobility

INTRODUCTION

Thermoplastic aromatic polyimides (PIs) are high performance engineering polymers that are used in a wide range of specialty applications and have been studied extensively using dynamic mechanical and dielectric analysis. The wide possibilities for changing the chemical structure of the diamine and dianhydride components of the thermoplastic aromatic PIs allow one to change their physical properties. In this work, the dielectric spectroscopy method is used to study the dielectric relaxation and molecular mobility in five thermoplastic aromatic PIs of different structures.

EXPERIMENTAL SECTION

Dielectric spectra were measured in the frequency range from 10^{-1} Hz to 10^{6} Hz and the temperature range from -100 °C to +400 °C (in increments of 10 °C and accuracy of 0.5 °C) using "Concept 81" spectrometer (Novocontrol Technologies GmbH & Co). The experimental data included impedance (imaginary and real parts) measurements for the cell with the sample [1].

RESULTS AND DISCUSSION

For all samples, the dielectric spectra, $\varepsilon^{//} = \varphi(f)$, show three regions of $\varepsilon^{//}_{max}$ due to relaxation of dipole polarization: the γ and β (in the glassy state) and the α (in the highly elastic state). At temperatures exceeding the α process, we observed a region of $\varepsilon^{//}_{max}$ corresponding to a conductivity relaxation.

The analysis of dielectric spectra was performed by describing complex permittivity ε^{*} with the empirical Havriliak-Negami (HN) equation (1). Relaxation time values for all processes at different temperatures calculated according to the HN equation testify that the $-\log\tau_{max}=\varphi(1/T)$ dependences for the γ and β process are linear and can be described by the Arrhenius equation which is typical of the local forms of molecular mobility. For the α process, dependences of $-\log\tau_{max}=\varphi(1/T)$ are nonlinear, which is characteristic of cooperative types of molecular mobility and its were reliably described by the empirical Vogel–Tamman–Fulcher (VTF) equation (2).

$$\varepsilon^{*}(\omega) - \varepsilon_{\infty} = \sum_{k=1}^{n} \mathrm{Im}\left[\frac{\Delta\varepsilon_{k}}{\{1+(i\omega\tau_{HN_{k}})^{\alpha_{k}}\}^{\beta_{k}}}\right] \quad (1) \qquad \tau_{max} = \tau_{0}\exp\left(\frac{B}{T-T_{0}}\right) \quad (2)$$

Comparing the data obtained earlier for linear thermotropic polyesters and PI of different structures, it can be assumed that the appearance of the β process reflects the limited rotations of the phenyl rings and the adjacent polar –O– groups. Parameters of the Arrhenius equation, $\log\tau_0$ и E_a, are ~20 s and ~32 kcal/mole, respectively, exceed those for local processes, which can be explained by some influence of intermolecular interactions. For the γ process, $\log\tau_0$ и E_a, are ~15 s ~11 kcal/mole, respectively, are typical for the local mobility mode. Concerning the molecular mechanism of the γ process, there is no unequivocal explanation. For the γ and β processes, the molecular mobility practically does not depend on the chemical structure.

For the α process, influence of chemical structures on the molecular mobility was estimated by value of glass transition temperature T_g. For linear polymers, T_g is mainly determined by intermolecular interactions between macromolecules. In PI, the main contribution to intermolecular interactions is made by the dispersion forces between the flat phenylene rings. The chemical structure of the investigated PI is selected in such a way that it is possible to determine the contribution of each element entering into the diamine and dianhydride component on T_g of different systems. So, the introduction of oxygen "bridges" –O– between phenylene rings into the diamine part makes macromolecule more flexible, the intermolecular interactions between macromolecules decrease, and as a result, decrease in T_g by 30^0C. T_g is also influenced by dipole-dipole interactions. It is shown that the replacement of the polar group $-O-$ by $-SO_2-$ in the dianhydride part of the macromolecule gives a decrease in T_g by 40^0C. This means that a consistent change in the chemical structure of the thermoplastic aromatic PIs, allows us to establish a correlation between the important characteristics as T_g and dielectric relaxation.

REFERENCES

1) Nikonorova N.A.; Didenko A.L.; Kudryavtsev V.V.; Castro R.A.: Journal of Non-Crystalline Solids, Vol. 447, (2016), p. 117.

Analysis of micro- and nano-pores of oil shale and oil shale ash

*M.Yu. Nazarenko, N.K. Kondrasheva, S.N. Saltykova**

max.nazarenko@mail.ru

Department of Processing mineral raw materials, Mining university, Saint-Petersburg, Russia

ABSTRACT

In this study, we investigated such properties of oil shale and oil shale ash as the total porosity, the value of the square of the specific surface distribution of micro-and nano-pores. The change in the total porosity of the oil shale occurs in four stages. .I n the first stage (25–200°C), the porosity increases on account of the liberation of water, carbon dioxide, and hydrogen sulfide. In the second stage (200–400°C), the porosity declines, on account of the bituminization of the shale. In the third stage (400–600°C) and the fourth stage (600–900°C), the porosity increases on account of semi coking and the decomposition of carbonate minerals. The average diameter of the pores of oil shale is in the limit of 8,962 nm - 21,244 nm.

Keywords: oil shale, oil shale ash, pores, porousness, surface area, micropore, sorbents

INTRODUCTION

Oil shale contain kerogen—that is, transformed organic matter from higher plants and simple organisms. They differ from coal in that they form a large quantity of tar on semicoking (20–70%, in terms of the organic component) [1]. After processing oil shales, a large quantity of oil shale ash remains (up to 50 wt % of the initial shale); its storage consumes enormous areas [2]. Oil shale also provide less energy, mass for mass, than other fossil fuels, including petroleum and natural gas. Nevertheless, interest in processing oil shale has risen sharply in the last decade. Researchers assert that effective industrial use of oil shales demands the use of not only the organic component but also the mineral component, so as to reduce the waste generation. [1-2]. The composition and surface reactivity of oil shales and spent shales were studied in [3-4]. On that basis, it seems that oil shales and oil shale ash may be used as sorbents, since they resemble materials such as sand and zeolite in terms of their composition (content of calcium, silicon, an aluminum oxides).

29

One of the important properties of the sorbents is the porosity and the specific surface area. On this basis, the aim of this work is the study of the porosity of oil shale and oil shale ash.

EXPERIMENTAL SECTION

We study oil shales from the Baltic Basin in the Leningradsk field (< 0,125 mm fraction) and also oil shale ash (<0,125 mm fraction).

The change in oil shale total porosity on pyrolysis in a nitrogen atmosphere and in air (at 25–900°C) is studied pycnometrically. A 2g sample is placed in the pycnometer, into which ethyl alcohol is poured. The sample is held at 25°C for 120 min.

The specific surface area and micro- and nano- pores of oil shale and oil shale ash were conducted on the analyzer, specific surface NOVA3200e. The cell device was weighed on an analytical balance. The mass was recorded. The cuvette was placed in samples weighing 2 g. In cuvettes with samples were prepared by vacuum and was heated to 1000C and aging at this temperature is 2 hours, thus, the pores of the samples were freed from possible moisture. After that, the cuvettes with samples were weighed, the weight recorded. The cuvette was again placed in the device, they created a vacuum and the cell was dipped in liquid nitrogen. When establishing temperature equilibrium between the liquid nitrogen and the cuvette with the sample cuvette was starting to dispensed gas is nitrogen, and with the help of pressure sensors recorded the absorption of nitrogen by the sample. Determination of the specific surface area was by the method of BET (Bruner-Emmett-teller) based on the fact that the solid surface is considered as a set of adsorption sites, and in a state of dynamic equilibrium speed condensing at the free space of the molecules of the gas phase equals the rate of evaporation of molecules from the occupied places.

CONCLUSION

Definitely that the total porosity of shale is 24%, oil shale ash is 38%. The change in the total porosity of the oil shale occurs in four stages.

1.I n the first stage (25–200°C), the porosity increases on account of the liberation of water, carbon dioxide, and hydrogen sulfide.

2. In the second stage (200–400°C), the porosity declines, on account of the bituminization of the shale.

3. In the third stage (400–600°C) and the fourth stage (600–900°C), the porosity increases on account of semicoking and the decomposition of carbonate minerals.

Porosity of oil shale after heat treatment in air atmosphere than in the nitrogen atmosphere, due to the oxidizing atmosphere and the consequent formation of a larger amount of gaseous products. The increase of porosity at temperatures above 800-900°C. With is due to the formation of a large number of silicates. For example, in the conversion of alpha-quartz into alpha-cristobalite volume increases of 15.7%, which provides the change in porosity.

It is established that oil shale ash (51,796 sm^2/g) has a specific surface area higher than that of oil shale (9,427 sm^2/g), which is mainly due to the fact that during heat processing, a large amount of volatile components thereby increasing the porosity.

The average diameter of the pores of oil shale is in the limit of 8,962 nm - 21,244 nm.

REFERENCES

Leimbi Merike R., Tiina H., Eneli L., Rein K. Composition and properties of oil shale ash concrete / Oil Shale. 2014. Vol. 31, №. 2. 147–160.

Bityukova L., Motler R. Composition of oil shale ashes from pulverized firing and circulating fluidized-bed boiler in Narva thermal power plants / Oil shale. 2010. vol. 27. № 4. 339–353.

Swift T., Mayer S. Study of thermal conversion of oil shale under N_2 and CO_2 atmospheres / Oil shale. 2010. vol. 27. № 4. 309–320.

Nazarenko M.Yu., Bazhin V.Yu., Saltykova S.N., Konovalov G.V. Physicochemical properties of fuel shale / *Coke and Chemistry*. 2014. №3. 44-49;

Nazarenko M.Yu., Kondrasheva N.K., Saltykova S.N. Surface reactivity of fuel shales from the Baltic basin/ *Coke and Chemistry*. 2016. №5. 33-37.

Structure and Sorption Properties of Clay Minerals

Vyacheslav N. Pak[*], *Yu. Yu. Gavronskaya, E. S. Ivanova*

Herzen State Pedagogical University of Russia, Saint-Petersburg

pakviacheslav@mail.ru

ABSTRACT

Comprehensive study of the composition, structure, and sorption properties of a typical representative of clay minerals has been performed. The efficiency of extraction of Ni^{2+} ions from aqueous solutions with a concentration of 10^{-5} to 10^{-2} M has been determined. Such sorbent characteristics as chemical and mineralogical composition, specific surface, porosity, maximum sorption capacity, and number of active sites have been found to be mutually consistent.

Keywords: clay minerals, phase composition, surface, active centers, porous structure, ion

INTRODUCTION

Ion-exchange/sorption properties of natural clay minerals are determined by a combination of interrelated factors, such as chemical and mineralogical composition, crystallinity, number of active sites, porosity parameters, and conditions and results of chemical and thermal treatment. However, studies on sorption properties of particular samples often do not consider the complete set of the aforesaid factors, especially in adsorption–structural studies, substantiation of the solid/liquid phase ratio, and independent estimation of the number of surface active sites by different methods. As a result, comparison of published data and prediction of sorption properties for particular samples become difficult [1-3]. The above stated also applies to some extent to solid surface studies. Nevertheless, fairly comprehensive and consistent description and analysis of interrelations of the above properties may be possible within the scope of conventional research practice.

MATERIALS AND METHODS

Typical clay samples of Russia north-west region were investigated. X-ray fluorescence analysis was performed with Shimadzu-13PK Instrument. Water adsorption was used to determine specific surface area and pore volume of clays. Sorption capacity of the samples towards Ni2+ was investigated as the decrease of ions content in solutions by the probes analysis with the use of EDTA.

RESULTS AND DISCUSSION

According to the X-ray fluorescence analysis data, the main components of the typical clay sample of Russia north-west region are the following oxides (wt %): SiO_2 (55.16), Al_2O_3 (15.52), CaO (6.51), Fe_2O_3 (5.33), MgO (2.73), K_2O (1.0), Na_2O (0.83), TiO_2 (0.81). This composition suggests potential sorption capacity the clay. The ratio [Al]/[Si] is about 1/3, which is optimal for ensuring required number and strength of surface active centers. In the examined sample, these centers are occupied mainly by Na^+ and K^+ ions; efficient exchange of the latter for heavy metal ions that form poorly soluble aluminosilicates in acid medium is quite possible. Therefore, the maximum sorption capacity of the clay may be estimated *a priori*. The overall concentration of Na_2O and K_2O in the clay sample is ~0.24 mmol/g, this value reflecting the maximum sorption capacity for doubly charged heavy metal cations. The major crystalline mineral component is montmorillonite while kaolinite and muscovite are concomitant phases. The presence of montmorillonite as the major phase (60%) is consistent with the [Al]/[Si] ratio equal to ~1/3 which reflects high potential ion-exchange capacity of the clay.

It is important to compare the experimental maximum sorption capacity for Ni^{2+} ($Q = 0.11$ mmol/g) – table 1 - with the prediction made on the basis of the sorbent composition (0.24 mmol/g; see above). Presumably, accessible active surface of the sorbent contains approximately a half of the total amount of alkali metal ions, and just those active sites on the surface involved in ion exchange process determine the observed sorption capacity. As a result, the degree of extraction of Ni^{2+} is 95–97% at low Ni^{2+} concentration in solution ($c = 10^{-5}–10^{-4}$ M).

Table 1 demonstrates that maximum sorption capacity $Q = 0.11$ mmol/g (0.22 mequiv/g) is consistent with the water monolayer capacity ($a_m = 0.24$ mmol/g). It looks as very specific and means that surface active sites accessible to water vapor are almost equally accessible to nickel ions in solution, which may be regarded as an indirect support for high degree of occupation of the sorbent surface by Ni^{2+} ions. In addition, estimation of the adsorption cross section of Ni^{2+} for $Q = 0.11$ mmol/g as $\omega = S_{sp} Q^{-1} N_A = 0.2$ (nm^2) matches well with the hydrated nickel ion size.

Table 1 Sorption parameters of Ni^{2+} for three types of clay of the northwest region of Russia

	1	2	3
Q, mmol/g	0.11	0.25	0.31
mequiv/g	0.22	0.50	0.62
ΔG, kJ/mol	−12.6	−13.5	−16.5
S_{sp}, m^2/g	14	30.5	36
Q,	**0.016**	**0.016**	**0.017**

mequiv/m^2			
$a_{m,}$ mmol/m^2	**0.017**	**0.017**	**0.017**

In general, limits of sorption in relation to Ni^{2+} ions are in good agreement with the change in free energy ΔG. It is important to note that the sequence is determined by the totality of sorbent properties, including the important role of the specific surface S_{sp}. Revealing, in this regard, are the results of casting the limit sorption values Q and capacity of monolayer of adsorbed water a_m (table). Convergence of values obtained from normalization per m^2 testifies to a single nature and nearest sorption/exchange surface activity of the investigated natural samples.

Finally, the results of potentiometric titration of an aqueous suspension of the sorbents are very representative. The initial pH value equal to 9.65 reflects equilibration between the glass electrode and alkaline surface of solid particles. The amount of acid necessary to adjust the suspension to a neutral pH value is 0.22 ± 0.03 mmol/g, which coincides with the maximum sorption capacity for Ni^{2+} ($Q = 0.22$ mequiv/g) and is close to the monolayer capacity of adsorbed water ($a_m = 0.24$ mmol/g). The observed agreement between independently determined parameters indicates high surface coverage of the clay sorbent with alkali metal ions and reflects their activity in both ion exchange and adsorption processes.

The main determinant of the set unity is the proximity of clay sorbents module-ratio, namely, Al/Si ~ 1/3.

ACKNOWLEDGEMENTS

This work was financially supported by the Ministry of Education and Science of the Russian Federation in the framework of the basic part of the State contract.

REFERENCES

[1] Bhattacharyya K.G., Gupta S.S. Advances in Colloid and Interface Science, 140, (1), (2008) 114-131.

[2] Kumrič K.R., Duk č A.B., Trtič-Petrovič T.M., Vucelič N.S., Stojanovič Z., Grbovič-Novacovič J.D., Matovič L.L. Industrial and Engineering Chemistry Research, 52 (2013) 7930-7939.

[3] Ivanova E.S., Gavroskaya Yu.Yu. Russian Journal of General Chemistry, 84 (2), (2014) 181-184.

Exfoliated Graphite From Graphite Nitrate Cointercalation Compounds: Production And Some Applications

A.A. Voitash[1], V.Yu. Vishnevsky[1], Yu.V. Berestneva[1], E.V. Raksha[1],
A.V. Muratov[1], A.B. Eresko[1], V.A. Glazunova[2], V.V. Burhovetskiy[2], G.K.
Volkova[2], M.V. Savoskin[1]*

*elenaraksha411@gmail.com

[1]L.M. Litvinenko Institute of Physical Organic and Coal Chemistry, Donetsk

[2]Donetsk Institute for Physics and Engineering named after A.A. Galkin, Donetsk

ABSTRACT

A triple graphite intercalation compound – graphite nitrate cointercalated with acetic acid and ethyl formate was used as a starting material for exfoliated graphite production. Carbon nanoparticles were obtained by sonication of the exfoliated graphite. Sorption capacity of the exfoliated graphite to benzene, engine oil and heavy oil was determined. Sorption of the Pd catalysts by exfoliated graphite from the cross-coupling reaction after target product extraction was studied.

Keywords: Exfoliation of graphite, graphite intercalation compounds, carbon nanoparticles, sorption capacity, cross-coupling reaction.

INTRODUCTION

Exfoliation of graphite is usually performed by the intercalation of graphite to form a graphite intercalation compound (GIC), followed by heating of the GIC. Recently we obtained new graphite cointercalation compounds via additional intercalation of different organic substances into the graphite nitrate [1, 2]. It was revealed that variation in the organic cointercalants nature and composition allows one to control the stability/lability of the resulting compounds within a wide range [1, 2]. This paper presents the results on the exfoliated graphite production starting from triple graphite nitrate cointercalation compounds (GNCC) as well as some applications of the exfoliated graphite for carbon nanoparticles generation, and water purification from ecotoxicants.

RESULTS AND DISCUSSION

A triple graphite intercalation compound – graphite nitrate cointercalated with acetic acid and ethyl formate was obtained as a starting material for exfoliated graphite production. The overall scheme of the process involves graphite nitrate synthesis as a first stage followed by cointercalation with organic substances. The resulted GNCC dried at 20 °C then undergoes thermal shock at 900 °C to obtain the exfoliated graphite. The obtained GNCC is black-gray powder resembling initial graphite (Figure 1a) but some exfoliated flakes (Figure 1b) are present in the sample even after drying at 20°C. Thus, this compound was expected to be perspective for nanoparticles [2] as well as exfoliated graphite production.

a   b   c

Figure 1 - SEM images of initial graphite (a), graphite nitrate cointercalated with acetic acid and ethyl formate (b), and of exfoliated graphite (c) obtained from GNCC

Samples of exfoliated graphite were sonicated (42 kHz, 70 W) in the ethyl alcohol for 1 hour to obtain carbon nanoparticles. The microstructure of obtained particles was investigated by transmission electron microscopy (TEM). It should be noted that started GNCC samples were compounds of the IV and II intercalation stages thus it was difficult to expect a complete graphite lattice exfoliation into monolayer graphene particles. TEM images of the obtained carbon nanoparticles are shown on Figure 2.

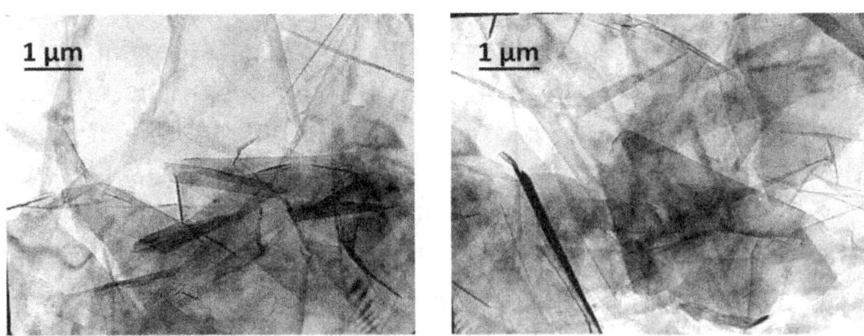

Figure 2. TEM images of carbon nanoparticles obtained by sonication of the exfoliated graphite

Exfoliated graphite has excellent sorption properties to some organic liquids, heavy oil or engine oil floating on the water surface. The sorption capacity of the exfoliated graphite to benzene, engine oil and heavy oil was determined. It was 71 g, 43 g, and 62 g per 1 g of the exfoliated graphite respectively.

As well, exfoliated graphite is promising sorbent for noble metals. The sorption of the Pd catalysts by obtained exfoliated graphite from the cross-coupling reaction after target product extraction was studied. We use 2-chloro-3-phenyl-pyridine synthesis (Figure 3) as an example of cross-coupling reaction that successfully applied in the large-scale pharmaceutical industry. SEM images of exfoliated graphite after Pd sorption are presented on Figure 3. Further investigation on the application of the obtained exfoliated graphite with Pd will be carried out.

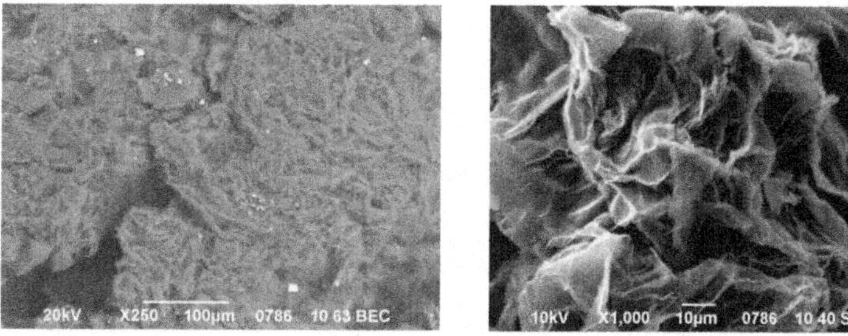

Figure 3. Scheme of the model reaction of 3-chloro-3-phenyl-pyridine synthesis and SEM images of exfoliated graphite after Pd sorption.

REFERENCES

[1] Savoskin, M.V.; Yaroshenko, A.P.; Whyman, G.E.; Mysyk, R.D. *J. Phys. Chem. Sol.* 2006, *vol. 67*, 1127 – 1131.

[2] Savoskin, M.V., Vdovichenko, A.N., Raksha, E.V., Berestneva, Yu.V., Vishnevskij, V.Yu., Padun, A.P., Verbenko, I.A., Yurasov Yu.I. Production of carbon nanoparticles by exfoliation of graphite nitrate and its co-intercalation compounds. In: *Proceedings of the 2016 International conference on "Physics, mechanics of new materials and their applications"*. Ed.: Parinov, I.A., Chang, Sh.H., Jani, M.A. - Nova science Publishers, Inc, New York, 2017. – 794 p. – P. 115-120.

Sol-Gel Synthesis, Research And Application Of Powders – Multiferroics With The Structure Of A 'Core-Shell'

Olga A. Shilova[1,2*], *Tamara V. Khamova*[1], *Gennady P. Kopitsa*[1,3], *Maxim M. Sychev*[1,2]

[1]Institute of Silicate Chemistry of the Russian Academy of Sciences, St. Petersburg, Russia, olgashilova@bk.ru

[2]St. Petersburg State Technological Institute (Technical University), Russia

[3]National Research Centre Kurchatov Institute: Konstantinov Petersburg Nuclear Physics Institute, Gatchina, Russia

ABSTRACT

New composite powders of $BaTiO_3$ with a sol-gel-derived 'skin-layer' on their surface have been prepared. The data of SEM, XRD, X-ray spectral mapping, SAXS, powder neutron diffraction, small-angle scattering of polarized neutrons, low-temperature adsorption of N_2, magnetic characteristics of composite powders and dielectric parameters of prepared composites (polymer/powder) have been submitted.

Keywords: powders, core-shell, powder mesostructure, fractal structure, organic-inorganic composite materials, magnetic properties, dielectric properties

INTRODUCTION

The multiferroics have both magnetic and ferroelectric properties, as well as a magnetoelectric effect due to the interaction of these components. This gives new opportunities in the design of electronic equipment.

Composite powders with a 'core-shell' structure: $BaTiO_3$ (core) /$CoFe_2O_4$ (shell) and more often in reverse order $CoFe_2O_4$ (core) /$BaTiO_3$ (shell) are successfully produced by co-precipitation and sol-gel processing [1, 2]. However, the mesostructure of the composites obtained, as well as the "synthesis conditions – structure – properties" dependencies, have not been adequately studied.

In the present work the results of studying the structure and properties of composite powders of the "ferroelectric (core) // magnetic shell" type are considered.

Moreover these composite powders have been used as fillers in a polymer matrix. Composites: "polymer matrix /// multiferroics (filler)" are prospective for use in electronic engineering as film capacitors. A number of complementary modern research methods have been used.

EXPERIMENTAL SECTION

The 'skin-layers' have been formed during treatment of a commercial powder $BaTiO_3$ by tetraethoxisilane-derived sols containing compounds of cobalt and iron. The synthesized powders, so called multiferroics, were used as functional fillers for dielectric composites (polymer/filler) based on cyanoethyl ether of polyvinyl alcohol (CEEPVA).

The structure, composition and magnetic characteristics of composite powders and dielectric parameters of prepared composites (polymer/powder) have been investigated using the following methods: scanning electron microscopy (SEM), X-ray diffraction (XRD), X-ray spectral mapping, small-angle X-ray scattering (SAXS), powder neutron diffraction, small-angle scattering of polarized neutrons (SAPNS), low-temperature adsorption of N_2 and dynamic light scattering in the laser Doppler anemometer mode.

RESULTS AND DISCUSSION

According to SEM, XRD and X-ray spectral analysis, the silica 'skin-layer' containing uniformly distributed nanoparticles of cobalt spinel ferrite ($CoFe_2O_4$) was formed on the surface of the powder. The specific surface of $BaTiO_3$ powder with a 'skin-layer' increased in ~ 17 times. This phenomenon can be associated with the formation on the powder surface of a more or less uniform layer, which changed the surface morphology. To confirm this hypothesis, X-ray spectral mapping of $BaTiO_3$ particles with a 'skin-layer' was carried out. It was shown that Si, Co and Fe are uniformly distributed over the surface of the $BaTiO_3$ particles. On the basis of the ζ-potential measurements the following assumption is made: the formation of 'skin-layer' takes place through the mechanism of electrostatic interactions of the oppositely charged powder particles and the sol particles ($\zeta_{BaTiO3} = -0,41$, and $\zeta_{sol} = +2.57$ at pH ~2).

The magnetic substructure of the powders with the structure of the "core ($BaTiO_3$)-shell ($SiO_2/CoFe_2O_4$)" was characterized by SAPNS. Estimations of the characteristic size R of magnetic-nuclear cross-correlations were obtained from the analysis of the patterns of magnetic-nuclear neutron scattering. The larger particles $CoFe_2O_4$ ~20-50 were observed using SEM. The magnetization curves for obtained materials were studied. The synthesized powders, so called multiferroics, were used as functional fillers for dielectric composites (polymer/filler) based on cyanoethyl ether of polyvinyl alcohol (CEEPVA).

The films based on the organic-inorganic composites (CEEPVA///BaTiO$_3$//SiO$_2$/CoFe$_2$O$_4$) were formed under the influence of an external magnetic field. It has been found that these organic-inorganic composites based on the powders with a 'skin-layer', especially, which were dried under the influence of a magnetic field, have a higher permittivity ε as compared to the reference sample (CEEPVA///BaTiO$_3$) based on the initial powder [3].

CONCLUSION

A powder with a core-shell structure is prepared on the basis of commercial BaTiO$_3$ by treated with a doped TEOS-derived sol.

The 'shell' material consists of an amorphous silica matrix having a surface fractal structure formed by silica particles of 25-50 nm in size depending on the processing temperature. Particles of CoFe$_2$O$_4$ with a characteristic average size of 30-45 nm are statistically uniformly distributed in the silica matrix. At a temperature of 1000 °C, the silica matrix begins to crystallize.

The ferroelectric BaTiO$_3$ powder with a magnetic SiO$_2$ thin 'shell', containing nanoparticles of CoFe$_2$O$_4$, is magnetically hard material. It is used as filler in a polymer matrix made of cyanoethyl ether of polyvinyl alcohol to produce a composite material for film capacitors.

The use of a multiferroic composite powder of BaTiO$_3$ ('core') and SiO$_2$/CoFe$_2$O$_4$ ("shell") as filler as well as an external perpendicular directional magnetic field during the drying of a "multiferroic //polymer "composite greatly increases the dielectric constant ε of the composite material. These composites can be used as a material for embedded capacitors and other electronic devices, as well as sensors, shields of electromagnetic radiation, etc.

REFERENCES

1.   Schileo, G., Recent developments in ceramic multiferroic composites based on core/shell and other heterostructures obtained by sol–gel routes (2013) Progr. Solid State Chem. (41) 4: 87-98.

2.   Selvi, M. M., Manimuthu, P., Kumar, K. S., Venkateswaran, C., Magnetodielectric properties of CoFe$_2$O$_4$–BaTiO$_3$ core–shell nanocomposite (2014) J. Magnetism Magnetic Mater. 369: 155-161.

3.   Sychov, M., Nakanishi, Y., Vasina, E., Eruzin, A., Mjakin, S., Khamova, T., Shilova, O., Mimura, H. Core-shell approach to control acid-base properties of surface of dielectric and permittivity of its composite (2015) Chem. Lett. (44) 2: 197–199.

Conrol Of Current And Voltage Oscillations In A Short DC Discharge Making Use Of External Auxiliary Electrode

A.S. Mustafaev[1], A.Y. Grabovskiy[1], V.I. Demidov[2], I. Kaganovich[3], M.E. Koepke[2]*

*Vladimir.Demidov@mail.wvu.edu

[1]St. Petersburg Mining University

[2]West Virginia University, Morgantown, West Virginia, USA

[3]Princeton Plasma Physics Laboratory, Princeton, New Jersey, USA

ABSTRACT

A dc discharge with a hot cathode is subject to current and voltage plasma oscillations, which have deleterious effects on its operation. In this paper, we demonstrate how it is possible to achieve reliable suppression of the discharge oscillations by installing an auxiliary electrode, placed outside of anode. By collecting a modest current through a small opening in anode, we show that the discharge becomes stable, in a certain pressure range. This method of avoiding current oscillations can be used for wide range of applications.

Keywords: plasma chemistry, dc discharge, electron distribution function, plasma instabilities negative resistance, heated cathode

INTRODUCTION

Gas-discharge plasma devices are widely used in plasma applications [1]. A better understanding of the physics behind these devices has been developed over the last few decades is made possible through development of new diagnostic techniques [2] and has allowed for the creation of discharge plasmas with more controllable parameters, including modification of the charged particle densities, temperatures (average energies), and electron energy distribution functions (EEDF). However, improving the performance of these new plasma tools is still a significant challenge for plasma engineering. Exploitation of nonlocal plasma properties allows additional dimensions and flexibility in adjusting plasma parameters. A remarkable property of such plasmas is that changing conditions in one place may lead to unexpected changes far away in another part of the plasma. One example of such a device with nonlocal plasma properties is a short dc discharge (several millimeters in length at the pressure of a few Torr, 10–100 μm for atmospheric pressure)

RESULTS AND DISCUSSION

It is known [3] that dc electric discharges can be unstable with respect to excitation of various types of oscillations and instabilities. Oscillations are excited due to ionization plasma instability related to the falling volt-ampere characteristic of the discharge or part of it. These oscillations are affected by external electric circuits. There are many regimes of oscillations, which can be identified. While plasma instabilities, in principle, can be harnessed for some purposes, for example, for the generation of voltage oscillations, they are harmful for many other applications, for instance, for the development of current and voltage stabilization devices, and, in this case, they should be suppressed. In this paper, we demonstrate how it is possible to achieve reliable suppression of the discharge oscillations by placement of the external auxiliary electrode outside the discharge and subtraction of the anode current through a small opening in the anode.

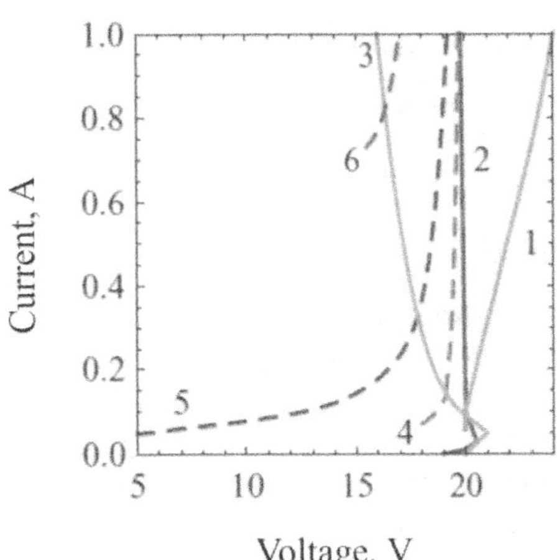

The discharge current-voltage characteristics for helium gas pressure of 0.6 Torr (1), 1 Torr (2), and 4 Torr (3). The diode regime is shown by solid curves (no current to the auxiliary electrode). Regimes with the auxiliary electrode are shown by the dashed curves for helium gas pressure of 1 Torr and different currents to the auxiliary electrode (0.1 A (4) and 0.4 A (5)). Curve (6) is for helium gas pressure of 4 Torr and current to the auxiliary electrode of 0.1

In figure the typical current-voltage experimental characteristics of the cathode-anode gap are presented. Curves 1, 2 and 3 correspond to the case without collection of the current by the auxiliary electrode. It is possible to see that in the diode regime, the discharge has a positive discharge differential resistance for the pressures below 1 Torr (curve 1). Increasing the gas pressure to 1 Torr leads to transformation of the discharge differential resistance to the slightly negative one (curve 2). Further increase of the gas pressure yields higher negative discharge differential resistance (curve 3). The presence of the negative resistance may lead to plasma instabilities and oscillations of the discharge voltage and current.

The collection of the anode current to the auxiliary electrode leads to a transformation of the discharge differential resistance from the negative to the positive one, as evident from curves

4, 5, and 6. Curves 4 and 5 were obtained for the helium gas pressure of 1 Torr and curve 6 is for 4 Torr. Each curve was obtained with fixed auxiliary electrode current. The experimental data show that an increase in the current drawn to the auxiliary electrode makes discharge differential resistance even more positive (compare curves 4 and 5). In all cases with the positive differential resistance, the oscillations are suppressed and are practically absent.

REFERENCES

1. A.S. Mustafaev, A.Y. Grabovskiy. High Temperature. 2017. Vol.55. No1. P.20-26.

2. A.S. Mustafaev, V.S. Soukhomlinov. Journal of Mining Institute. 2016. Vol. 222. P.864-868.

3. A.S. Mustafaev. Technical Physics. 2001. Vol.46. №4. P.472-483.

\\

Effets Of Graphite Intercalation With Cesium In A Thermoionic Energy Converter

A.S. Mustafaev, V.S. Sukhomlinov*

alexmustafaev@yandex.ru

St. Petersburg Mining University, St. Petersburg, Russia

ABSTRACT

The thermionic energy converter (TEC) with inter-electrode low-temperature plasma and cesium vapor dynamic flow in the inter-electrode gap (IEG) demonstrated an increase in efficiency of up to 20%. This was mainly reached due to the decrease in the effective electron work function to an anomalously low value of the order of 1 eV from a perforated nickel collector covered with nanosized graphite flakes under the converter working conditions. SEM X-ray microanalysis of the collector surface layers was performed, and the model of the effects proposed. In a such kind of collector of TEC with a nanosized graphite coating, multilayered intercalation of Cs atoms in the graphite/graphene surface layers was found during work in cesium vapors at a pressure of ~1–10 Torr. This probably explains the anomalously low electron work function and the high efficiency of TEC.

Keywords: thermionic energy converter, electron work function, nickel collector, nanosized graphite, elemental microanalysis, graphite intercalation by cesium atoms

INTRODUCTION

Composite carbon nanostructures with a low work function are of considerable interest for numerous applications ranging from electron emission devices to ion-battery technologies. It is known that adding hydrogen atoms or alkali metals to the surface of carbon structures can lead to remarkable reduction of the surface work function. These surface effects can be used in new thermionic energy converters (TEC) of the thermal power into electric.

Studies on TEC performed at the Institute for Physics and Power Engineering and aimed at seeking methods for drastically raising the efficiency (several-fold) and operating life (by approximately an order of magnitude) relative to the previously achieved level (~10% and ~1 year, respectively) in TOPAZ and Yenisei space nuclear power units [1] showed that it is possible to achieve high efficiency of converters with electrodes of advanced materials [2-4].

RESULTS AND DISCUSSION

As is well known, the electromotive force developed by TEC is mainly determined by the electron work function (EWF) of the emitter material, and the output voltage losses depend on the EWF of the collector material and plasma voltage losses in the interelectrode gap (IEG). The efficiency of TEC can be increased mainly by decreasing the collector EFW. Earlier [5, 6], high efficiency of laboratory TEC was demonstrated experimentally by continuously circulating cesium vapors through a nickel collector with many holes with a diameter of 0.1 mm. The effect was observed if the collector surface was coated with graphite using an aquadag suspension, and not observed on a metal surface.

The experiments performed in [5, 6] were repeated independently on an improved laboratory TEC with a perforated collector with a coating of nanosized graphite flakes. As TEC electrodes, we used an emitter of vacuum-melted polycrystalline Mo with a diameter of 14 mm and a thickness of 11 mm with a Pt (~3 μm) coating and a collector (with a diameter of 8 mm) of laser-perforated nickel foil with a thickness of 0.2 mm (121 holes with a diameter of 0.1 mm on a square of 4×4 mm). The emitter was heated by electron bombardment; T_E = 1350 K, T_C = 750 K.

The panoramic SEM images of the collector surface show concentric rings around each hole in the collector (figure). They evidently suggest that diffusion occurs on the collector surface from the boundary of the hole.

Then the distribution of elements between the neighboring holes was measured in different regions of the collector. Spectrum 12 from the region of the concentric diffusion ring in the figure showed Ni, Mo, C, Cs, and O lines.

During the operation of TEC, molybdenum evaporated from the emitter and fell on the collector. In spectrum 7, there were no Cs and O X-ray lines

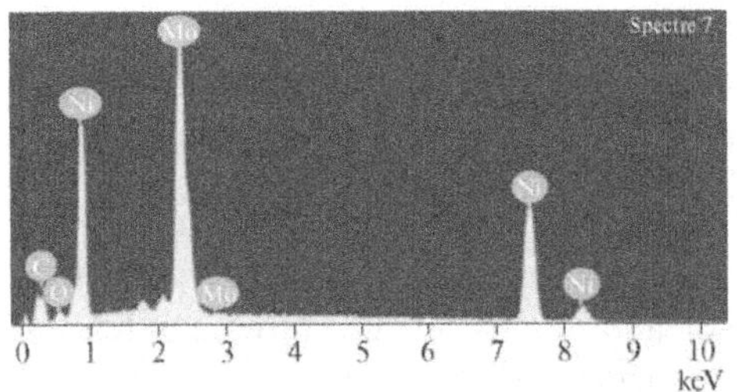

SEM image and X-ray fluorescence spectra of the surface of a perforated nickel collector coated with nanosized graphite after prolonged exposure to Cs vapors.

outside the limits of the concentric diffusion ring in the figure. Nearly equal partial concentrations of Cs and O atoms were observed at different points of the diffusion rings around the collector holes, but not observed outside the holes. This may be explained by diffusion (intercalation) of these atoms from the hole boundaries into surrounding multistoreyed (with a height of ~1 μm) layers of nanosized graphite-graphene. The observed Ni concentration inside the rings is lower than outside because Cs and O intercalated into multilayered graphite-graphene screened the lower-lying Ni atoms of the substrate from fast electrons.

REFERENCES

1. G. M. Gryaznov and V. Ya. Pupko, Priroda (Moscow, Russ. Fed.), No. 10, 29 (1991).

2. V. I. Yarygin, J. Clust. Sci. 23, 77 (2012).

3. V.I.Yarygin, A.S. Mustafaev, Russian Journal of Physical Chemistry B, V.9, No. 4, p. 546 (2015).

4. A.S. Mustafaev et al., Russian Journal of Physical Chemistry B, V.36, No.2, p.94 (2017).

5. L. Holmlid, in Proceedings of the Thermionic Energy Conversion Specialist Conference, Goteborg, Sweden, 1993, p. 47.

6. R. Svensson, L. Holmlid, and E. Kennel, in Proceedings of the Thermionic Energy Conversion Specialist Conference, Goteborg, Sweden, 1993, p. 93.

Investigation and formation of a sorption gas sensor based on carbon nanostructures with complex relief.

A.A.Rezvan and V.S.Klimin*

South federal university, Taganrog, Russia

arezvan@sfedu.ru

ABSTRACT

In this paper design and a laboratory-technological instruction for the manufacture of a sorption gas sensor with a sensitive element based on an array of vertically oriented carbon nanostructures have been developed. The essence of this technology lies in the production of graphene films by the method of plasma-chemical deposition from the gas phase on the surface of vertically-oriented nickel nanostructures.

Keywords: nanotechnology, graphene, carbon nanostructures, sorption sensors of gas, plasma chemical vapor deposition (PECVD).

INTRODUCTION

At present, various researchers associate further prospects for the development of nanotechnologies with carbon-based structures. Such close attention and wide interest of science carbon nanomaterials were caused primarily by the discovery of fullerenes, carbon nanotubes and graphene, which have unique physical and mechanical properties. Thus, due to its box structure, resistance to ionizing radiation and a hexagonal two-dimensional monatomic crystal lattice, which makes it possible to determine small concentrations of adsorbed gas, carbon nanostructures are one of the promising materials of micro- and nanoelectronics. In this connection, development and creation of portable gas sensors on their basis is actual.

EXPERIMENTAL SECTION

In the course of this work an experimental study of forming a sorption gas sensor using vertically oriented carbon nanostructures method was carried out. Experimantal layout is a small insulating substrate on which measuring electrodes, a sensor element and a heater are located on opposite sides of the substrate. The heater is necessary for degassing, restoring the sensitive element and maintaining a constant temperature of the processes occurring on the surface of the semiconductor during gas chemisorption.

So, initially, Experimental layouts were chemically purified subtrates Si(100), on which silicon dioxide Si_3N_4 films were synthesized by the method of plasma chemical vapor deposition(PECVD). Also, using the magnetron sputtering method, a catalytic Ni layer was formed on this structure. On the obtained structure, nickel nanowires were formed by a combination of methods of focussed ion beams and plasma chemical etch. Carbon nanostructures on the profiled surface of this experimental model were obtained by plasma-chemical deposition from the gas phase, the carbon-containing gas was acetylene C_2H_2. To the finished model, over the sensitive layer, electrical contacts connected to the signal processing device were connected.

RESULTS AND DISCUSSION

As a result of the experimental studies it was determined that the made model of the resistive gas sensor based on the sensing element consisting of a carbon nanosystem is susceptible to acceptor molecules of gas NO2 and donor molecules of gases NH3, CO, H2O vapor with different degrees of sensitivity depending on the concentration n.

REFERENCES

Ageev O.A., Klimin V.S., Solodovnik M.S., Eskov A.V., Krasnoborodko S.Y. The study of influence of the gas flow rate to etched layer thickness, and roughness of the anisotropy field of gallium arsenide is etched in the plasma chemical etching process. Journal of Physics: Conference Series, 2016, T. 741, № 1. p. 012178.

Klimin V.S., Solodovnik M.S., Smirnov V.A., Eskov A.V., Tominov R.V., Ageev O.A. A study of the vertical walls and the surface roughness GaAs after the operation in the combined plasma etching. Proceedings of SPIE - The International Society for Optical Engineering "International Conference on Micro- and Nano-Electronics 2016", 2016, p. 102241Z.

Ilina M.V., Blinov Y.F., Ilin O.I., Klimin V.S., Ageev O.A. Resistive switching of vertically aligned carbon nanotube by a compressive strain. Proceedings of SPIE - The International Society for Optical Engineering "International Conference on Micro- and Nano-Electronics 2016", 2016, p. 102240U.

Investigation of the influence of the modification processes of nanostructures microrelief based on silicon dioxide films by the method of plasma-chemical etching

V.S.Klimin, A.A.Rezvan, I.N.Kots*

South federal university, Taganrog, Russia

arezvan@sfedu.ru

ABSTRACT

In this paper using of plasma chemical etch method was examined for implementation of profiling processes of surface layers of structures based on silicon dioxide SiO_2, which can be used for further epitaxial growth. The essence of the technology is rational sequence of methods of focused ion beams and plasma chemical etch(PCE). Preproduction models were formed with unique surface relief, which depends on doses of embedded ions. Also there was determined the range of dose values of the Ga + ion beam for silicon doping.

Keywords: silicon dioxide, local ion doping, plasma chemistry, focused ion beam, nanotechnology, semiconductor structures.

INTRODUCTION

For future micro- and nanoelectronics plasma chemical etch or PCE of semiconductor structures is is one of the main methods for synthesizing nanoscale structures with sufficiently low linear dimensions less than 100 nm. This method for the formation of etching process uses chemical active particles, which are generated in a low-temperature plasma of a high-frequency (HF) discharge. Thanks to which it is possible to control changes in the concentrations of electrons and ions in the discharge, which can significantly affect the demand for the characteristics of the etching process. So, PCE is characterized by high resolution, selectivity and anisotropy of etching. In connection with this, the choice of the method of chemical etching from the high-frequency (HF) plasma discharge for the surface modification processes is actual.

EXPERIMENTAL SECTION

In this paper an experimental study of the effect of parameters of the processes of modifying and surface treatment of structures based on silicon dioxide films by the method of plasma chemical etching. Experimental layouts were chemically purified subtrates Si(100), on which silicon dioxide SiO2 films were synthesized by the method of plasma chemical vapor deposition(PECVD).

Further local doping of silicon dioxide regions with gallium ions Ga^+ was carried out.on the surface of obtained structures by the method of focused ion beams. This process was conducted on a scanning electron microscope "Nova NanoLab 600" with ion column and scanning probe microscope (SPM), with the help of which the samples were scanned. The order of the beam movement was determined by the required sequence of coordinates in the template file. The beam propagation time varies from 100 to 1000 ms. This pattern file is executed with apertures with a beam current of 1, 10, 30 pA with a beam diameter of 7, 12, 16 nm, respectively, and an ion beam energy of 30 keV in the absence of ion beam overlap at neighboring points of impact.

There selective parameters of plasmachemical etching processes for each mask were used for the formation of nanoscale structures based on silicon dioxide films. Experimental layouts were subjected to PCE in a high-frequency inductively-coupled plasma in a " STE ICP E68" installation in several stages.The first stage was the etching of silicon dioxide films of experimental samples, the working gas mixture of the process being SF_6. The second stage was final, the structures were etched in a gas mixture of BCl3 to remove the remaining Ga particles.

The determining parameters of the silicon dioxide PCE were chosen so that the chemical etching mechanism predominated over physical spraying. With the selectivity condition, volumetric nanoscale structures were obtained with the following parameters of PCE: RF power supplied to the source inductor, 200 W; capacity of the capacitive pair supplied to the substrate, 10 W; the pressure in the chamber is 1.33 Pa; the bias voltage is 40 V, the working gas flow is $N_{SF6} = 15$ cm^3 / s. To obtain the etching depth of silicon 9.73 and 252.1 nm with these parameters, the etching time is 30 and 120 s, respectively. The bias voltage is necessary to accelerate the etching process, but it was chosen in such a way that the etching rate of silicon was significantly greater than the etching rate of the gallium mask. Further investigations of the surface structure were carried out by scanning electron microscopy and probe nanotechnology

RESULTS AND DISCUSSION

After the completion of a number of experimental studies, the dependence of the geometric parameters of the etched layer on the etching time was obtained, and an analysis was made of the rates of the etching process for various gas flows. In addition, the effect of gas flows on the roughness of the etched surface was studied and it was shown that the determining parameter of the angle deviation from the vertical is the gas flow and the crystallographic direction of these structures.

REFERENCES

V.S. Klimin, M.S. Solodovnik, V.A. Smirnov, A.V. Eskov, R.V. Tominov, O.A. Ageev et al,, Proceedings of SPIE - The International Society for Optical Engineering, 102241Z (2016)

V.S. Klimin, M.S. Solodovnik, A.V. Eskov, S.Y. Krasnoborodko et al., Journal of Physics: Conference Series, 741, 012178 (2016)

Ageev O.A., Klimin V.S., Solodovnik M.S., Eskov A.V., Krasnoborodko S.Y. The study of influence of the gas flow rate to etched layer thickness, and roughness of the anisotropy field of gallium. arsenide is etched in the plasma chemical etching process. Journal of Physics: Conference Series, 2016, T. 741, № 1. p. 012178.

Atomic force microscopy of nanocomposites based on zinc oxide with different additives

Evgeniya V. Maraeva[1], Vyacheslav A. Moshnikov[1],*

Nadezhda D. Yakusheva[2], Igor A. Pronin[2] and Igor A. Averin[2]

*jenvmar@mail.ru

[1] Saint Petersburg Electrotechnical University "LETI"

5 Prof. Popova str., Saint-Petersburg, 197376, Russian Federation

[2]Penza State University

40 Krasnaya str, Penza, 440026, Russian Federation

ABSTRACT

The article is concerned with the investigation of nanomaterials based on ZnO with different additives (Fe, Cu). A correlation between technological modes of production (annealing temperature, composition) and the values of the fractal dimension of the samples is under discussed.

Keywords: Porous nanoparticles, atomic force microscopy, fractal dimension, zinc oxide

INTRODUCTION

Currently, the nanocomposites based on metal oxides are widely used for creation gas sensors with a percolation structure [1-3]. The devices based on such structures allow one to reach extremely high values of gas-sensitive [4]. One of the most important ways to study such structures is fractal analysis based on data getting with the use of atomic force microscopy (AFM). There are different approaches to estimate the fractal dimension of the nanostructures, such as method of cube counting, triangulation method, power spectrum method etc. However, all of these methods depend on many factors (for example the size of scan area). All of them can be implemented in the graphic program «Gwyddion» for scanning probe microscopy (SPM) data analysis.

EXPERIMENTAL SECTION

The main purpose of this work was to study the peculiarities of the surface relief in the nanocomposites based on zinc oxide and to determine the influence of annealing temperature and composition of investigated materials on the value of their fractal dimension. The stated purpose implies the following objectives:

1. To study the nanocomposites based on zinc oxide with different additives (Fe, Cu) with the use of scanning probe microscopy;

2. To identify the relationship between technological modes of production (composition, annealing temperature) and the values of fractal dimension of the samples.

The study of surface morphology of nanocomposites based on zinc oxide were carried out by means of scanning probe microscopy with atomic - force microscope NTEGRA (NT-MDT, Russia) with the use of silicon probes NSG with a radius of 10 nm in tapping mode. Evaluation of the fractal dimension was carried out in a graphics program for SPM data analysis «Gwyddion». The methods of cube counting and triangulation method based on the analysis of AFM images were used.

RESULTS AND DISCUSSION

For research purposes, a set of samples based on zinc oxide with additives of Fe and Cu, heated at different temperatures (420 K, 870 K) was obtained [4].

With the use of atomic force microscopy method it was found that all the samples were characterized by the presence of close-packed grain system. From the morphological point of view the layers without additives of Cu and Fe were uniform despite of the processing conditions, in all parts of the surface the same relief was observed. The range of heights at different areas of a surface was 800 – 1200 nm.

In some cases (the samples with Cu and Fe additives) the surface contained inclusions that represented the core with radiating channels. The surface areas free from impurities were characterized by typical samples relief.

For all the patterns the calculation of fractal dimension values was carried out with the use of the program «Gwyddion». The results are shown in table 1

Table 1 – The results of fractal analyses

Sample number	1a	2a	3a	1b	2b	3b
The	Z	Z	Z	Z	Z	Z

composition of the sample	nO	nO (Fe)	nO (Cu)	nO	nO (Fe)	nO (Cu)
Annealing temperature, K	870	870	870	420	420	420
Fractal dimension (triangulation method)	2,54	2,60	2,55	2,58	2,58	2,33
Fractal dimension (method of cube counting)	2,47	2,52	2,46	2,47	2,48	2,36

It can be seen from the table 1 that for ZnO (Cu) pattern which was heat-treated at T=420 K the surface structure was characterized by the lowest value of the fractal dimension.

CONCLUSIONS

Thus, the atomic force microscopy data analysis showed that the presence of alloying elements (Fe, Cu) primarily affects the distribution of grains on the surface of the zinc oxide nanocomposites. The variation of the heat treatment regimes in the range from 420 K to 870 K slightly affects the surface morphology of the samples. The lowest value of the fractal dimension characterizes the surface of ZnO (Cu) nanocomposites, annealed at T=420 K. For other types of the samples the fractal dimension varies slightly depending on the conditions of synthesis and depends mainly on the method of calculation.

ACKNOWLEDGEMENTS

The reported study was supported by the Ministry of Education and Science of the Russian Federation within the framework of the project № 16.897.2017/ПЧ.

REFERENCES

[1] Relationship between the photocatalytic and photoluminescence properties of zinc oxide doped with copper and manganese / Pronin I.A., Donkova B.V., Dimitrov D.Tz., Averin I.A., Pencheva J. A., Moshnikov V.A. // Semiconductors. 2014. vol. 48, № 7, p. 842–847.

[2] Metal oxide SnO_2 - ZnO - SiO_2 films prepared by sol-gel / Abrashova E.V., Kononova I.E., Moshnikov V.A. // Smart Nanocomposites. 2013. vol, 4, № 2, p. 1-7.

[3] Study of porous sol-gel nanocomposites based on silicon dioxide and tin dioxide modified by fullerenol $C_{60}(OH)_n$ (n = 22-24) / Maraeva E.V., Istomina M.S., Moshnikov V.A., Nalimova S.S., Maximov A.I., Alexeev N.I., Semenov K.N.// Journal of Physics: Conference Series. 2016. vol. 690, p. 012031.

[4] Theoretical and experimental investigations of ethanol vapour sensitive properties of junctions composed from produced by sol-gel technology pure and Fe-modified nanostructured ZnO thin films / Pronin I.A., Averin I.A., Yakushova N.D., Dimitrov D.T., Krasteva L.K., Papazova K.I., Chanachev A.S., Bojinova A.S., Georgieva A.T., Moshnikov V.A // Sensors and Actuators A: Physical. 2014. vol. 206, p. 88-96.

Non-Destructive Evaluation Of Strength Nanofeatures Of Constructional Materials

V.V. Nosov[1,2], I.V. Matviian[1], A. R. Yamilova[1]

[1]Saint-Petersburg Mining University, St. Petersburg, Russia,

[2]Peter The Great St. Petersburg Polytechnic University, St. Petersburg, Russia

ABSTRACT

Within period of manufacturing and exploitation of different materials there are could be observed complex physicochemical processes of interaction between their individual components and these components with environment. Revealing of such mechanisms helps to understand its distinct features and to optimize technological processes of manufacturing materials with certain characteristics. The solution could be based on interpretation of results of acoustic emission (AE) tests from the perspective of kinetic model of AE of heterogeneous materials.

Keywords: acoustic emission, heterogeneous materials, model of acoustic emission

INTRODUCTION

Research technique lies in experimental defining of effect of different technological and exploitation factors on quantities of parameters of AE model's coefficients.

RESULTS AND DISCUSSION

Presented model of timing dependence on number of AE impulses has the following view:

$$N_{\sum}(t) = V \iiint_{\Delta t, f, u} \Phi(\Delta t, f, u) du df d\Delta t \cdot C_0 \int_{\mu}^{\mu + \Delta \omega} \psi(\omega) \left\{ 1 - \exp\left[-\int_0^t dt' / \theta\left(U_0, \omega(t')\right) \right] \right\} d\omega \,, \qquad (1)$$

where $\theta(U_o, \omega(t')) = \tau_0 \exp\{[U_o - \gamma \sigma(t')]/(KT)\}$

Every parameter of the model [1] has its specific physical nature and depends on distinct factors what allow to reveal mechanisms of impact of these factors on material's features:

- parameter $V \iiint\limits_{\Delta t, f, u} \Phi(\Delta t, f, u) du df d\Delta t \cdot C_0 = k_{AE} C_0$ characterizes amount of

AE sources which are literally structural elements which we can be "hear" via AE equipment during the process of destruction;

- correspondence of the variables of $\Psi(\omega)$ function characterizes the degree of inhomogeneity of material's mechanical state at a molecular level;

- parameter U_o (activation energy of destruction process of molecular links) doesn't depend on state of material structure and is defined through characteristics of interatomic interaction (chemical ties) of structural element;

- parameter γ (activation volume) is characteristic of molecular nanostructure of material and it is faintly sensitive parameter to its chemical nature.

The point of the survey lay in defining of systematical variability of the parameters within changes of distinct technological and exploitation factors and was based on possibility of operational evaluation of these parameters due to results of acoustic emission tests.

Different samples of heterogeneous materials were exposed to destructive AE tests in regimen of steady loading with constant speed of tension rise. It was noticed during the tests that the view of timing dependences N_Σ (t) affected by such factors as size of filling material particles, time of isolation of the samples after their manufacturing, heat treatment and chemical saturation of uppermost layers, what can be explained by changing of degree of structural inhomogeneity of materials.

REFERENCES

[1] Nosov V.V. Diagnostics of machines and equipment. Textbook. Saint-Petersburg: Lan', 2016, p. 376 (in Russian).

[2] Nosov V.V., Matviyan I. V., Mechanics of non-uniform materials. Textbook. Saint-Petersburg: Lan', 2017, p. 276 (in Russian).

[3] Nosov V.V. Mekhaniks of composite materials. Laboratory and pracrical excersices. Textbook.. Saint-Petersburg: Lan', 2013, p. 240. (in Russian).

[4] Nosov V.V.. Defektoskopiya, 2014, № 12, 24-35. (in Russian).

[5] Nosov V.V. , Potapov A. I. Acoustic Emission Testing of the Strength of Metal Structures under Complex Loading //Russian Journal of Nondestructive Testing, 2015, Vol. 51, No. 1, pp. 50–58.

[6] Nosov V. V. On the Principles of Optimizing the Technologies of Acoustic-Emission Strength Control of Industrial Objects// Russian Journal of Nondestructive Testing,

2016, 2016, Vol. 52, No. 7, pp. 386–399.

[7] Nosov V. V., N.A. Zelenskii. Estimating the Strength of Welded Hull Elements of a Submersible Based on the Micromechanical Model of Temporal Dependences of Acoustic-Emission Parameters// Russian Journal of Nondestructive Testing, 2017, Vol. 53, No. 2, pp. 89–95.

[8] Nosov V. V. Acoustic-Emission Quality Control of Plastically Deformed Blanks// Russian Journal of Nondestructive Testing, 2017, Vol. 53, No. 5, pp. 368–377.

Modification of zinc films by chemical and physical methods

Vladimir V. Tomaev[1], Vladimir A. Polyshyk[2], Nikolai S. Pshchelko[3]**, Kirill L. Levine[1], Yuri I. Kuzmin[1]*

*tvaza@mail.ru

**nikolsp@mail.ru

[1]St. Petersburg Mining University, 199106, St. Petersburg, Russia

[2]St. Petersburg university of information technologies, mechanics and optics, Russia, 197101, St. Petersburg, Kronverkskuy pr. 49

[3]Military academy of communications named after S.M. Budenny St. Petersburg

ABSTRACT

Method of thermal treatment of zinc films in non-pressurised volume was developed and evaluated. Zinc films on glass substrate were subjected to thermal treatment in the presence of selenium vapors in air atmosphere, and combined thermal-electric field treatment. Morphology was assessed by SEM. Structural changes in initial and modified Zn films were discussed. Size and statistical parameters of zinc nanocrystals was determined. Was shown that by suggested method was possible to control zinc films morphology and composition in wide

Keywords: zinc films, surface morphology, nanocrystals of zinc and zinc selenide, selenium vapors, new phase nucleation.

INTRODUCTION

Zinc selenide (ZnSe) is known by its photo-resistive, photo- and electro-luminscent properties. Due to its high transparency in optical and infrared regions, it is used for manufacturing entry lenses and windows in optoelectronic devices [1].

Problem of obtaining, studying and usage of nanocrystalline films based on zinc halcogenides is on the edge of study [2-4].

In [5] it is realized the idea of optimization of photosensitive lead selenide (PbSe), by heating precursor films in the presence of selenium powder in air atmosphere in non-hermetically closed vessel. As a result of this procedure it was possible to obtain PbSe films with infrared sensitive capability at ambient temperatures (without cooling).

Polycrystalline ZnSe can be obtained by interaction of zinc with selenium vapors [5]. Monocrystals of cubic singonia were grown from vapor phase and melt, while monocrystals of hexagonal singonia – only from vapor phase. ZnSe films were obtained by thermal evaporation of the compound and condensation on substrate heated to 150 - 250 °C.

MATERIALS AND METHODS

Method in [5] describes chemical precipitation of PbSe from multicomponent mixtures under next following conditions:

- films are placed at non-hermetically close vessel, where certain ratio between surrounding volume and surface area of heating films is maintained, (usually ratio of total surface area to treating is 20-30.

- heating is carried out in the presence of metallic selenium.

Heating of Se in powder form allows saturating vessel with vapors of high partial pressure. Diffusion at temperatures of treatment creates conditions of controlled Se penetrating to lead. Varying experimental conditions allows varying Se concentration, decreases percentage of overoxidated sites and simultaneously decrease number of Se vacancies in PbSe crystalline lattice.

EXPERIMENTAL SECTION

Depositing selenium on zinc

In this paper same method as described above was applied to treat zinc films by Se vapors.

 Se surface-modified Zn films were obtained as a result of chemical interaction of Zn surface with Se vapors.

Izobaric conditions from one side, isolating from ambient atmosphere from the other, effectively decrease amount of oxygen during the reaction with simultaneous increasing of Se partial pressure. Oxygen becomes effectively consumed at initial stages of treatment and does not interfere Se deposition further. Oxygen partial pressure decreases $2.0 - 2.2$ times during the reaction, while Se vapors pressure competitively increases. Therefore, inn non-hermetically closed vessel, same effect as for hermetically closed is achieved, while complexity of experimental set-up is significantly decreased.

Considering high ionic contribution to chemical bonding, low thermal conductivity, high non-stoichiometry, polymorphic trend, high melting temperature, halcogenides synthesis by classical methods meets certain experimental difficulties [4].

Therefore, alternative halcogenide synthesis methods, especially with surface modified on nano-level, is of interest both from fundamental, and applied side.

In this paper it is suggested method of thermal treatment in selenide vapors. Applied to thin film coatings, this method allows creating nanometer-size selenide structures integrated to coating and chemically bonded to it.

Thermo-electric treatment of zinc

Zinc films of ~1000 nm thickness were obtained by vacuum thermal evaporation method on glass substrates (Fig. 1).

Fig. 1 Surface morphology of zinc films obtained by thermal evaporation onto glass substrate.

Fig. 2 Zinc crystals after treatment with selenide vapors.

For the majority of crystals, hexagonal structure is typical, which is indicative for rather high purity of grown Zn crystals and conditions of either homogeneous, either heterogeneous conditions of crystalline nucleation.

Cross-sectional Zn crystalline dimensions range from 200 to 1000 nm for films of nearly 1000 nm thickness.

Zinc crystals were heated to 250°C for 2 hours in the presence of selenide powder in non-hermetically closed vessel by method analogous to [6].

As a result of treatment in Se vapors, Zn crystalline substrate was coated by nanometer (~20-50 нм) size particles (Fig. 2).

Comparison of morphologies of untreated (Fig. 1) and treated (Fig. 2) in Se vapors Zn films shows that at treated film surface morphological features of a smaller size (20 – 50 nm) are formed, which were not present at initial film.

Also, at the edges of hexahedrons on perimeter protrusions can be noticed. This phenomena can be explained by interacting between Zn and Se vapors.

Another advantageous way of surface modification is developing by us method, where simultaneous temperature and electrical gradient in air atmosphere results in promoting electro-adhesion effects and creating films with significant differences in structure from ones synthesized without electrical field.

Zn film on glass was modified by constant electrical potential U=300 V, exposition time was 10 min and temperature 250°C. Positive potential wazsa applied to Zn. Precursor sample was placed in a holder that at the same time was upper electrode. Simultaneous effect of thermal treatment and electric field was studied as in [7,8]. Control sample was placed near treated sample, without applied voltage.

The influence of temperature and electric field on adhesion and phase composition of metallic zinc films was studied by experimental setup shown in Fig. 3. Experiments were performed in dry air atmosphere. Films for deposition were deposited on glass substrates.

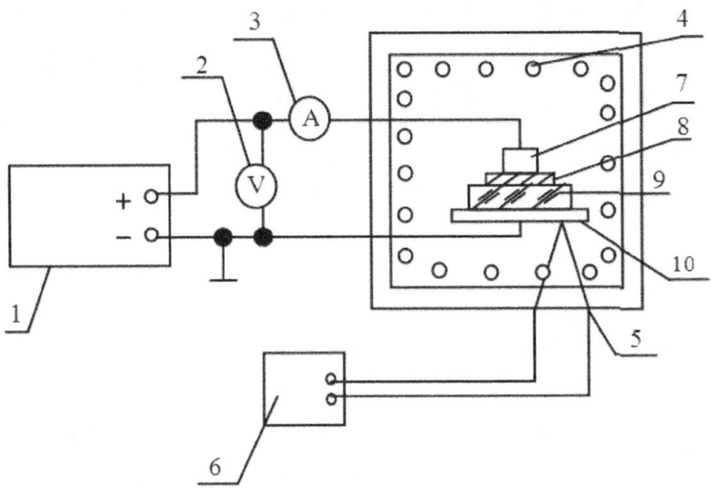

Fig.3 Experimental setup for electro-adhesion experiments. 1 – power supply. 2 – voltmeter, 3 – current recorder. 4. Oven, 5 – thermo-couple, 6 – high sensitive potential meter, 7, 10 electrodes, 8 – substrate for deposition, 9 – dielectric plate.

Treating sample with zinc film was hold in a device serving upper electrode at the same time. Thermal treatment with simultaneous action of electric field was studied. Control sample (outside of the electric field) was placed nearby. Modification of film at glass substrate was handled at U=300 V, treatment time was $t = 10$ min and $T = 250$ C°.

Obtained results show increased surface resistance ($R_1 \approx 15$ Ohm/sq) comparing to untreated sample ($R_2 \approx 4$ Ohm/sq). This is probably due to Zn oxidation by oxygen ions electro-migrating from glass to Zn. Transition from not-transparency to semi-transparency confirms this assumption. As known, ZnO are optically semi-transparent in visible band.

As a result of complex (temperature-electric) treatment, partial Zn oxidation to ZnO occurs. Large-gap (3.37 eV) semiconductor is a high-ohmic semiconductor with resistivity $\sim10^{-3}$ Ohm·cm) at room temperature. Partial Zn oxidation at applied experimental conditions results therefor in a significant decrease in conductivity,vwhich is for pure Zn $\sim5,9\cdot10^{-6}$ Ohm·cm).

Surface morphology was studied by scanning electron microscopy, while surface composition was evaluated by elemental analysis. Those measurements were carried out on Zeiss Merlin) electronic microscope.

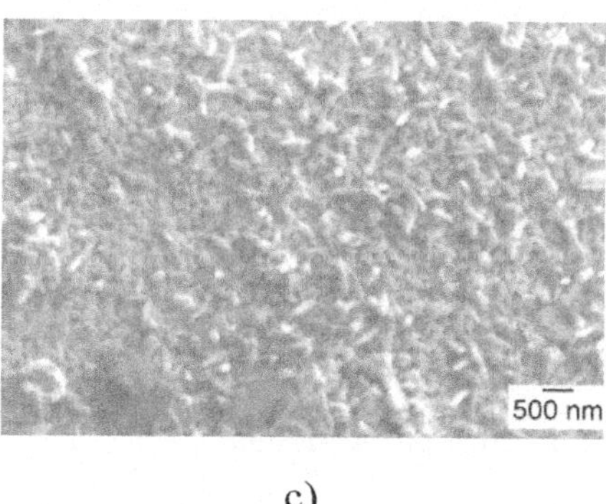

c)

Fig. shows microphotographs of films surfaces obtained at different stages of treatment.

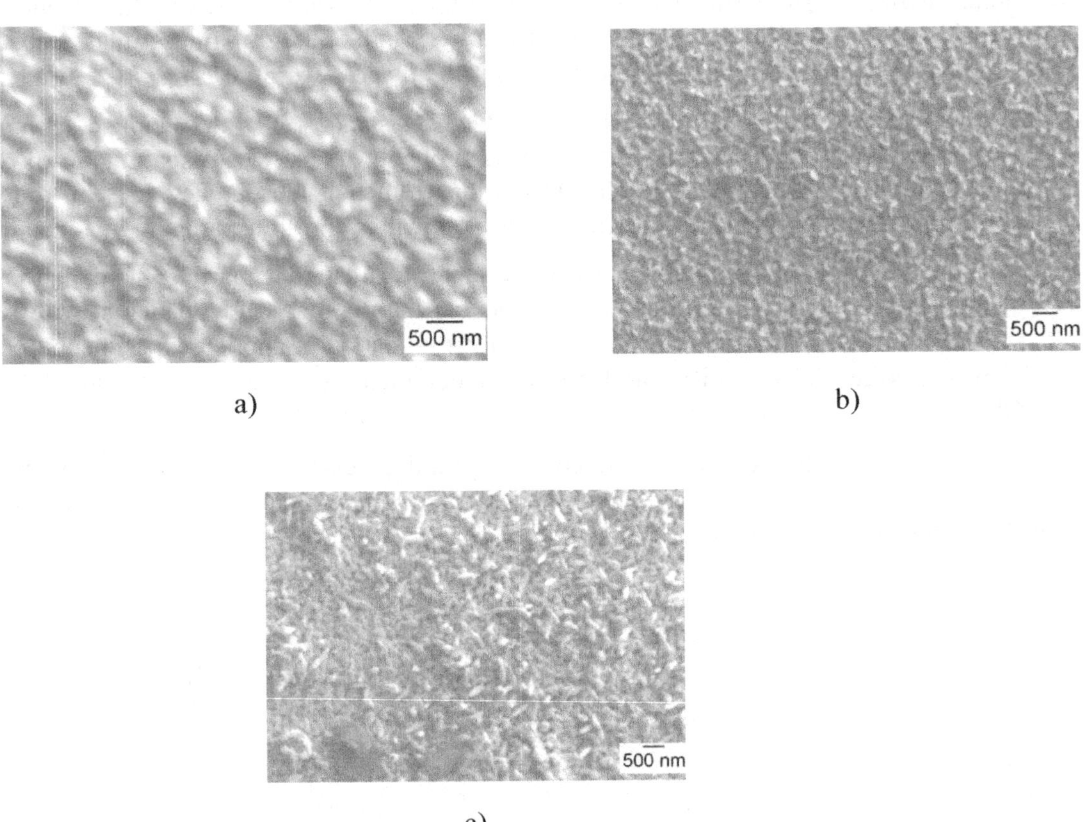

a)

b)

c)

Fig. 4. Treated and untreated films SEM images.

a) – precursor film, b) – film affected by thermal treatment in air, c) films thermally treated with applied cross-sectional electric field. Exposure time 10 min, temperature 250 C°

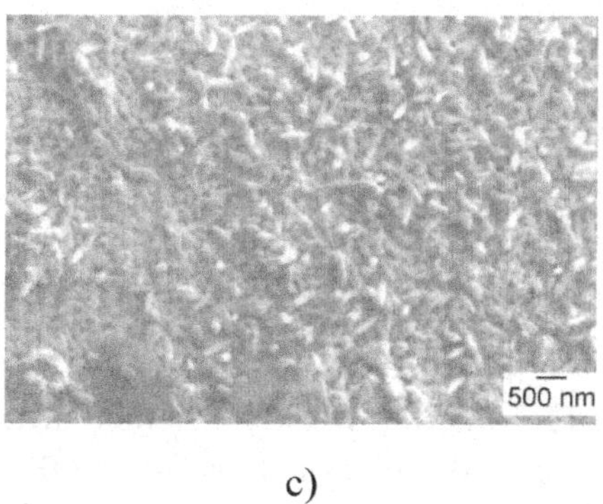

c)

Fig. shows that surface consists of separate, adjacent to each other crystals, which size ranges from 50 to 200 nm. From

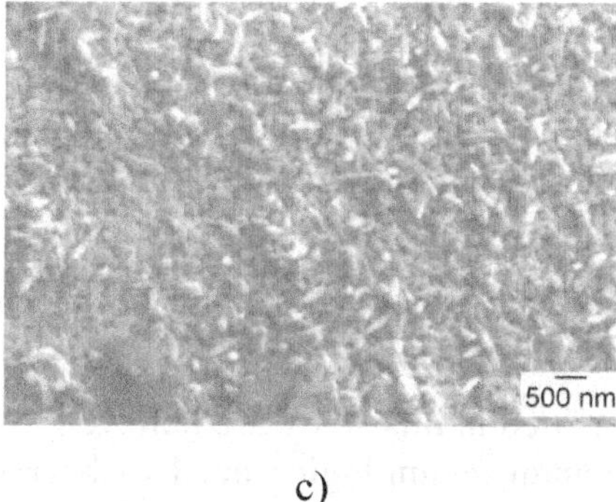

c)

Fig. b) it is clear that temperature does not strongly affect film structure, resulting only with its partial oxidation. Fragment of Zn film surface after 10 min temperature treatment in electric field (

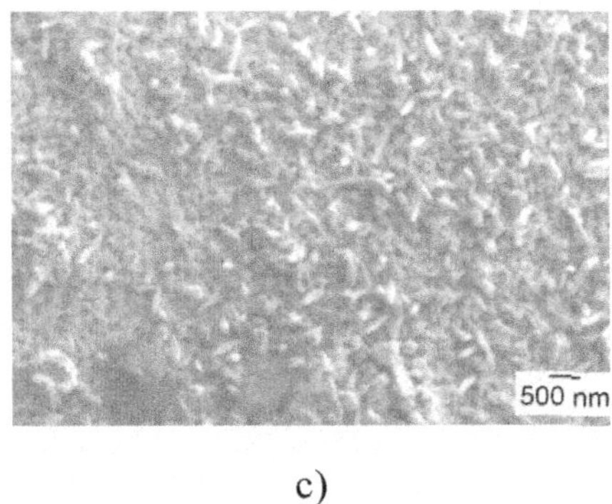

c)

Fig. c) is featured by large amount weakly oriented towards surface one-dimensional crystals. While cross-sectional size of those crystals is 50…100 nm, their length is 500… 1000 nm. The majority of filament-shape crystals are oriented towards surface randomly. Therefore, electric field significantly changes film structure: non-uniform electric field favors obtaining ZnO films with developed surface. Those ZnO films are composed from one-dimensional crystals of nanometer size.

CONCLUSIONS

Preliminary studies have shown that by varying technological regimes of thermal treatment under constant electric voltage, this is possible to alter concentration and linear dimensions of linear zinc oxide crystals.

Metal was converted to semiconductor by treating in the presence of oxygen and electric field.

Technological findings obtained in this study are believed to be useful for micro fabrication appliances, where expensive vacuum technologies need to be replaced by more cost-effective surface treatment methods.

REFERENCES

1. Panchal C.J., Opanasyuk A.S., Kosyak V.V., Desai M.S., Protsenko I.Yu. Structural and sub structural properties of the zinc and cadmium chalcogenides thin films // J. Nano-Electron. Phys. 3 (2011) No1, P. 274-301.

2. Dubey J.P., Tiwari R.K., Upadhyaya K.S., Pandey P.K. Crystal dynamics of zinc chalcogenides I: an application to ZnS // Turk J Phys (2015) 39: 242 – 253, http://journals.tubitak.gov.tr/physics/

3. Dubey J.P., Tiwari R.K., Upadhyaya K.S., Pandey P.K. Crystal dynamics of zinc chalcogenides II: An application to ZnSe // IOSR Journal of Applied Physics (IOSR-JAP) e-ISSN: 2278-4861.Volume 7, Issue 5 Ver. I (Sep. - Oct. 2015), PP 67-75 www.iosrjournals

4. Triboulet R. Growth of Zinc Chalcogenides / In book: Crystal Growth Technology, 2003, pp.497-523,.

5. Mironov M.P., Markov V.F., Mascaevfa L.N., Diakov V.F., Mukhamediarov R.D., Muhamedzianov H.N., Smirnova Z.I., Method o f sensibilization of chemically precipitated lead selenide films for IR radiation, Patyent RU 2357321, (2009).

6. JCPDS database No. 03-065-5973.

7. Pshchelko N. S., Sevryugina M.P. Modeling of physical and chemical processes of anodic bonding technology // Advanced Materials Research, Vol. 1040, 2014, p 513-518

8. Pshchelko N.S., Tomaev V. V. Polarization of Surface Layers of Ionic Dielectrics at the Interface Between the Electroadhesive Contact and a Dielectric // Glass Physics and Chemistry. Vol. 42 No.1. 2016 . P.105 -109.

Influence of the substrate on the formation of AgI films prepared by laser ablation

Vladimir V. Tomaev[1,2] , Tatiana V. Stoyanova[1]

tvaza@mail.ru

[1]Saint Petersburg Mining University, St. Petersburg, Russia

[2] Institute of Chemistry, Saint-Petersburg State University, St. Petersburg, Russia

ABSTRACT

AgI films with 1 μm thickness were synthesized by laser ablation method on substrates of different materials: cover glass, fiberglass and copper foil. The morphology of the surface of all silver iodide films was studied by scanning electron microscopy (measured with the Zeiss Merlin microscope), and their elemental composition was investigated by micro-X-ray spectral analysis. It was shown, that the morphology and composition of the substrate cause a significant effect on the formation of the morphology of AgI films.

Keywords: Silver iodide films, laser ablation method, substrates of cover glass, fiberglass and copper foil.

INTRODUCTION

To create a synthesis technology, it is very important to understand the mechanism of the process and the role of factors that determine the kinetics of the reactions. Since most of the processes are structurally sensitive, the most promising substrates for the development of nanotechnologies for the formation of nanoobjects and their use in model processes are monocrystals whose surface has a controlled structure.

Nevertheless, the formation of nanoobjects on substrates that do not have a definite crystal orientation (cover glass, fiberglass, copper foil) can also be very promising because at the early stage of nucleation of nuclei of a growing phase, the action of the substrate on the formation of nanocrystals of a polycrystalline film can be minimal and at this time "quasi-free" growth may occur.

EXPERIMENTAL SECTION

Sputtering of the films on copper and synthetic-resin bonded [SRB] paper laminate (getinax) substrates was carried out [1] by laser ablation [2] using an Xe-Cl excimer ultraviolet laser with a radiation wavelength of 30 nm, a pulses energy of 10 mJ, a duration of 30 ns, and repetition rate of 14 Hz. In previous work [3,4] was shown that this method allows to obtain AgI films with a composition that coincides with the composition of the target. In the process of growth, the shape of the surface (degree of roughness) can play an increasing role and at the final stage of growth the shape of the surface of the film can "copy" it.

It was observed a correlation between the morphology of the surface of the getinax and copper foil substrates with the corresponding morphologies on the surface of the AgI films. The surface of the silver iodide film deposited on the getinax repeats the morphology of the getinax substrate and has an inhomogeneous character. The pore size on the surface in some cases reaches several microns. However, the average size of individual grains is 100-200 nm.

To explain the results obtained, the table gives the coefficients of thermal expansion of all interacting surfaces: AgI; copper foil; Getinax. As can be seen from the table, the coefficient of thermal expansion of hexagonal β-AgI has a negative value for all temperatures at which this phase exists (from 0 to 423 K) [3]. At the same time, the coefficients of thermal expansion of copper and getinax have a positive coefficient of thermal expansion [4,5].

Table. Coefficients of thermal expansion of the film material (AgI - silver iodide) and substrates (copper foil and getinax)

Material	Coefficient of thermal expansion ($10^{-6}\,°C^{-1}$)	Reference
AgI	-8±4	[3]
Cu	16,5	[4]
Getinax	7,9-8,7	[5]

In this respect, the interaction of a relatively thin hexagonal β-AgI film with a negative coefficient of thermal expansion with thick substrate layers (copper and getinax), each of which has a positive coefficient of thermal expansion, can be very effective, which leads to a change in the surface of the film due to large difference coefficients of thermal expansion. A similar effect was previously observed by the authors on the example of other materials in interaction with β-

AgI [6].The analysis of the substrate effect on the surface morphology of the AgI film grown by laser ablation is based on the classical theory of nucleation [7] and the Gibbs-Volmer equation.

Explanation of the obtained results depending on the properties of the film, substrate and technological conditions is possible with the help of the mechanisms of Wolmer-Weber, Frank van der Merve and Stransky-Krastanov.

CONCLUSION

AgI films were synthesized by laser ablation method using an Xe-Cl excimer ultraviolet laser on substrates of different materials: cover glass, fiberglass and copper foil. The film surfaces were studied with scanning electron microscopy It was observed a correlation between the morphology of the surface of substrates with the corresponding morphologies of the surface of the AgI films. The analysis of the substrate effect on the surface morphology of the grown films is based on the classical theory of nucleation [7] and the Gibbs-Volmer equation. An explanation of the results obtained, depending on the properties of the film, substrate and technological conditions, is possible with the help of the Wolmer-Weber, Frank- van der Merve and Stransky-Krastanov.

The results of the analysis make it possible to conclude that it is possible to control the degree of nanodispersity of the silver iodide film surface by changing the substrate material and the technological application parameters. Such a modification, as was shown in [7], can lead to structural phase transitions of silver iodide, as well as a change in the grain size in the film, which as a result may lead to an increase in the conductivity with respect to Ag+ ions.

ACKNOWLEDGEMENTS

THIS WORK WAS PARTLY SUPPORTED BY RUSSIAN SCIENCE FOUNDATION 17-03-00121.

REFERENCES

1. S. Hull. Superionics: crystal structures and conduction processes. Rep. Prog. Phys. 2004, 67, pp 1233–1314.

2. D. Bäuerle. Laser Processing and Chemistry. –Berlin, Heidelberg: Springer-Verlag, 2000. p.788

3. M. Volmer and A. Weber. Keimbildung in Übersättigten Gebilden, Z. Phys. Chem. Leipzig 1926, pp 119 - 277.

4. Harvey G. and Fletcher N.H. Thermal expansion of β-silver iodide at low temperatures 1980, J. Phys. C: Solid St. Phys., 13 pp.2969-2973.

5. Koshkin N.I., Shirkevich M.G. Handbook of elementary physics. 1972, M. Science.. p.256

6. Getinaks electrotechnical sheet. Technical conditions. GOST 2718-74. 1997, M. Publishing house of standards.. p. 24

7. Tomaev V.V.,Tver'yanovich Y.S.,Bal'makov M.D. Methods of control over the phase composition of nanostructured silver iodide Russian Journal of Electochemistry, 2017, Volume 53, Issue 7, pp 777–781

Capabilities Of Marnetic Abrasive Finishing In The Field Of Nanotechnology

E.G. Zlotnikov, A.I. Keksin, L. E. Koroboyankwe

Saint Petersburg Mining University

zlotnik_evg@mail.ru keksin.a@mail.ru boysonjr@gmail.com

ABSTRACT

The article explains the advantages of magnetic abrasive treatment over the traditional methods used in the mechanical process of finishing. The possibility of forming nanorelief surfaces of various components and products is shown. Presented also, are the results of experimental studies on magnetic abrasive finishing of complex profile surfaces of cutting tools.

Keywords: magnetic abrasive finishing, nanorelief, nanotechnology, finishing methods.

GOALS

The aim of this article is to show capabilities of magnetic abrasive finishing in the field of nanotechnology.

INTRODUCTION

Since the last quarter of the twentieth century, research in the field of nanotechnology [1] has become a priority in Scientific research. Nanotechnology refers to technological processes and systems of machines and mechanisms, capable of performing ultra-precise operations on a scale of several nanometers [2]. It is known that micro- and nanorelief of machine components surface layers are formed at the final stages of the technological process of manufacturing parts, i.e. at the finish operations.

Currently, traditional methods of finishing (grinding, honing, lapping, etc.) do not allow for the provision of a high-quality level of the surface layer [3], which must be formed on many components in various spheres of human activity: in engineering (for example, in electronics, optics, lasers); in the ecology, medicine, etc. So, for example, the roughness quantity provided by honing and grinding on the surface of a component is $Ra = 25$ nm (0.025 µm), and $Ra = 13$ nm (0.013 µm) respectively [3], while one of the primary tasks of the electronics industry in the manufacturing of silicon monocrystal plates is to decrease the height of microroughness from 10 nm (0.01 µm) to 1 nm (0.001 µm) [4].

MATERIALS AND METHODS

Providing such a high-quality level of the components surface layer is possible by creating progressive technologies based on the use of physical and chemical sources of energy. One such source is the magnetic field energy, which is increasingly used in many fields of technology. Based on the use of magnetic field energy, a method of magnetic abrasive finishing has been developed, which makes it possible to form a nanorelief with unique properties [4].

Processing by this method consists of oriented as well as selective abrasive micro-cutting and micro-smoothing. The essence of this process is that, at relatively high values of micro-protrusions, the powder grains are contacted mainly with the shell peaks, which are concentrators of the magnetic field lines. Each working element (grain) in the magnetic field is set by the largest axis towards the surface to be treated. During wear and dulling of vertices, the element is reoriented in such a way that the newly formed largest axis is guided along the magnetic field lines. As a result, the surface of the workpiece is machined with sharp edges, i.e. the process of oriented abrasive cutting [6] takes place. This feature of magnetic abrasive finishing gives huge advantages over traditional methods of finishing operations. For example, magnetic abrasive finishing of silicon monocrystals plates, details of optics and laser technology allows for the formation of nanorelief surfaces with an unevenness height of less than 2 nm [4].

In the Mechanical department laboratory of machines with CNC, in St. Petersburg Mining University, experimental studies were carried out on magnetic abrasive finishing of complex profile surfaces of cutting tools. Machine Taps M16 × 2 made of high-speed steel   P6M5 were taken as samples. Used to form a magnetic abrasive powder based on titanium carbide and iron (TiC + Fe) was a magnetic abrasive brush in the workspace of the electromagnetic installation system based on a CNC milling machine.

The best result was obtained during magnetic abrasive finishing with powder granularity Δ = 160 μm, magnetic induction B = 0.6 T, polishing time t = 60 s. The parameter of the roughness of the rear surfaces of the cutting edges of the taps was achieved within Ra = 0.061 ... 0.09 μm (61 ... 90 nm) with the initial value Ra = 0.16 ... 0.22 μm [7, 8].

RESULTS AND DISCUSSION

With the use of magnetically abrasive powders of less granularity, the selection of the regime parameters and the increase in the finishing time of the surfaces of the cutting tools, it is possible to obtain a nanorelief with a roughening parameter of 40 ... 20 nm. Advantages of the method of magnetic abrasive polishing include the formation of a surface layer with favourable compressive stresses and increased microhardness, a significant reduction in the number of surface defects, high productivity and the possibility of process automation.

CONCLUSION

In conclusion it can be said that magnetic abrasive finishing has a tremendous technological potential. The unique combination of the combined action of magnetic field pulses on the material structure and the "soft" modes of removing the defective surface layer makes it possible to form a surface with a minimum of defects and high-performance properties, such as resistance to corrosion, wear and mechanical stress, by a magnetically abrasive method.

REFERENCES

[1] Shevchenko L.B. Information resources nanotechnology. – URL: http://cyberleninka.ru/article/n/informatsionnye-resursy-po-nanotehnologiyam (accessed date: 31.10.2016). (in Russian).

[2] Taniguchi N Current status in, and future trends of ultraprecision machining and ultrafine material processing. Annals of CIRP, Vol. 32/2, 1983, P. 573-582.

[3] Sunil Jha V. K. Jain. Nano-finishing techniques. – URL: http://web.iitd.ac.in/~suniljha/nanofinishing.pdf (accessed date: 31.10.2016).

[4] Khomich N.S. Magnetic-abrasive machining of the manufactured articles. – Minsk: BNTU, 2006. – 218 p. (in Russian).

[5] Kremen G.Z. Machining time estimation for magnetic abrasive processes. International Journal of Production Research, Vol. 32, 12, 1994, 2817-2825.

[6] Akulovich L.M. Fundamentals of magnetic abrasive finishing of metal surfaces. – Minsk: BGATU, 2012. – 316 p. (in Russian).

[7] Keksin A. I. Methods of increasing the quality of the thread pitches / A.I. Keksin, V.V. Maksarov // Agronomy Research. – 2013. – Vol. 11, 1. – P.139-146.

[8] Keksin A. I. Technological improvement of the quality of complex surfaces by the method of magnetic abrasive polishing / A.I. Keksin, V.V. Maksarov // Metalloobrabotka. - 2017. - 1 (97). - P. 47-57. (in Russian).

Reactivity of Solid Surfaces in the Chemical Nanotechnology

Yu. K. Ezhovskii

St. Petersburg State Technological Institute (Technical University),

St. Petersburg, 190013 Russia

ABSTRACT

The concept of the macromolecular structure of solids and quantitative estimation of surface reactivity using inductive constants is applied to the estimate of surface nanostructures on dispersed and single crystal matrices. The possibilities of the proposed approach are demonstrated by the example of the synthesis of oxide nanostructures on a silicon surface.

Keywords: reactivity of solid surfaces, dispersed and single crystal matrices, oxide nanostructures

INTRODUCTION

The development of the chemical nanotechnology of low dimensional systems on a solid surface are mainly associated with the possibility of predicting the conditions of reactions using quantitative estimation of their reactivity. One such approach is a correlation analysis system based on the principle of the linearity of free energies and is successfully used in the chemistry of high molecular weight compounds [1]. According to the core models of the structure of solid materials and the chemistry of supramolecular compounds developed on its basis [2, 3], any chemical transformations on a solid surface can be considered as similar to those of a polymer. In this case, an approach based on application of the principle of the linearity of free energies and a correlation-analysis system in the chemistry of a solid surface can become the basis for the creation of a quantitative theory of surface reactions.

EXPERIMENTAL SECTION

In accordance with the core concept of the structure of solids [2], any solid compound is a combination of a chemically inert core ($X–$) and reactive functional groups of the surface (Y) (e.g., –OH). All surface reactions with the participation of these groups not affecting the structure of the core can be considered as similar to polymer transformations [3]. The ability of the core as a substituent to any type of interaction is quantitatively characterized by the corresponding constant (σ_X), the numerical value of which has a relative character, i.e., makes it possible to estimate the effect of the given substituent with respect to the reference one. This makes it possible to apply the principle of the linearity of free energies for different solid compounds having similar-type functional groups within the formal approach and use linear correlations of the type:

$$A = A_{XY} + \alpha_Y \sigma_X, \qquad 1$$

where A is any correlated quantity (e.g., the velocity constant (K) or log$K$); α - is the proportionality coefficient reflecting the sensitivity of the reaction center to the effect of the substituent in the given reaction. Indices Y and X refer, respectively, to the reaction center (the functional group) and the solid core.

The influence of the core of the solid, and, more exactly, the surface group of its atoms can be characterized by the value of the inductive constant (σ_S), which can be determined, e.g., from shifts of the frequencies of valence vibrations of the element–hydride bond, in particular, the Si–H bond in the functional group of the surface compound. This bond is very sensitive to the effect of substituents and the shift of its frequencies almost excludes the effect of steric factors. In this case Eq. (1) takes the form:

$$v_i = v_0 + \alpha \left(\Sigma \, \sigma_i + \sigma_\Sigma \right), \qquad 2$$

where v_i, v_0 are the frequencies of the valence vibrations in the given compound and in the reference one, e.g., methyl-substituted silane or siloxane.

RESULTS AND DISCUSSION

The coefficients of Eq. (2) for a reaction series of solid oxides were determined using samples of silicon dioxide (aerosil A-175) with different chemosorbed chlorinated silanes. As a result of mathematical processing of the experimental data (v_i) obtained for a series of surface compounds and values of Taft constants ( ) for such substituents as Cl_2MeSiH ($\Sigma\sigma_i^* = 2.88$), Cl_3SiH ($\Sigma\sigma_i^* = 5.76$), Cl_2PhSiH ($\Sigma\sigma_i^* = 3.48$) and $ClPh_2SiH$ (($\Sigma\sigma_i^* = 1.10$), we determined the coefficients of the relation (2) and obtained the expression:

$$v_i = 2064 + 19.21 \left(\Sigma \, \sigma_i^* + \sigma_* \Sigma \right), \qquad 3$$

Analysis of experimental results on determining the reactivity of hydroxyl groups of silica, silicon and arsenide–gallium matrices made it possible to assume that the hydroxyl groups on the real surface of a semiconductor matrix will possess an activity determined by the prehistory of the sample, namely, the thickness of the oxide layer formed after preliminary treatment. It is possible to determine this influence from the inductive effect and reveal the quantitative regularity of the reactivity of OH-groups of the surface of the crystal on the thickness of the oxide layer. The obtained data made it possible to establish the quantitative relationship between the thickness of the oxide layer on the silicon matrix and its inductive effect, and, consequently, the activity of hydroxyl groups. The values of the inductive constants make it possible to quantitatively estimate the effect of the solid core and the state of its surface at different temperatures, predict the direction of surface reactions with different reagents and establish their stoichiometry, when the use of direct chemical-analytical methods is difficult.

REFERENCES

[1]. V. A. Pal'm, Bases of Quantitative Theory of Organic Reactions (Khimiya, Leningrad, 1977) [in Russian].

[2]. V. B. Aleskovskii, Zh. Prikl. Khim. **55**, (1982). 725-730.

[3]. V. B. Aleskovskii. Course of Chemistry of Supramolecular Compounds (SPb. Gos. Univ., St._Petersburg, 1996) [in Russian].

[4]. Yu. K. Ezhovskii. Russ. Chem. Rev. **73**, (2004). 195- 204.

Influence Of The Material Structure Of Instrument On The Quality Of Processing While Milling With Cutting Ceramics

A.D. Khalimonenko, R.R. Rakhmankulov, I.V. Gorshkov*

Saint Petersburg Mining University

khalim76@rambler.ru

ABSTRACT

The paper deals with the problems that arise when milling the mounting surfaces with a tool equipped with cutting ceramics, based on an analysis of the requirements applicable to the quality of the surface layer and modern technologies for its providing. The results of the research showed that the quality improvement in the case of finishing milling depends on the operating capacity of the ceramic tool, determined by the parameters of its microstructure.

Keywords: Cutting ceramics, operating capacity of tool, milling, processing quality, mounting surface.

GOALS

The purpose of this work is to determine the operating capacity of a ceramic tool that affects the quality of machining while milling the precise elements of machine parts. The operating capacity of the ceramic material is determined on the basis of its microstructural parameters.

INTRODUCTION

A feature of modern production is the use of new technologies and tool materials, which, having high cutting properties, ensure, at the same time, the necessary quality of processing of the precise elements of machine parts. Such precise machine parts include elements of the mounting surfaces, which are subject to relatively high requirements to ensure the necessary dimensional accuracy, surface roughness, the accuracy requirements for the geometrical elements and their mutual disposition.

The most optimal is the technology: roughing, semi-finished and fine (fine) milling, the latter being recommended for use with cutters equipped with replaceable plates of cutting ceramics, which provide the necessary quality of processing at the highest productivity. This technology has a disadvantage - low stability of cutting ceramics with intermittent cutting, leading to early failure and loss of efficiency of not only a separate cutting insert of the mill, but also the entire tool.

If one of the ceramic plates fails to work, the load on the remaining plates increases, the tool quickly fails, which immediately leads to loss of processing quality. Hence it follows that the task of this study is to determine the operability of each ceramic plate in order to install in the milling cutters of such ceramic plates whose performance is at the same level.

MATERIALS AND METHODS

Cutting ceramics differs from other materials with high strength, reduced brittleness and tendency to cracking, while possessing high cutting properties [1]. The operating capacity of the cutting ceramics is influenced by such microstructure parameters as the amount of carbide grains and their average diameter. Ceramic plates having a finer grain structure (average grain size of 1.2 ... 1.5 µm) and a smaller percentage of porosity (about 8%) have greater operating capacity compared to ceramic plates in which the structure consists of large-diameter grains (medium grain size 1.9 ... 2.5 µm) and there is a relatively large percentage of porosity (about 12%) [1, 2].

The selection of ceramic plates for milling, according to their performance, was carried out by the correlation dependence of the microstructural parameters and the specific electrical resistance of the tool material. Ceramic plates with a larger microstructure have values of resistivity close to R = 10 Ω, and ceramic plates with a smaller microstructure have electrical resistivity parameters close to R = 100 Ω [1]. A similar range of parameters of the electrical resistivity of ceramic plates does allow them to be in the same group in terms of their microstructural parameters and in terms of operating capacity [1, 3].

For the experiments, the end mill was used and the recommended finishing modes adopted for the standard technology for the preparation of mounting surfaces were selected.

RESULTS AND DISCUSSION

A series of experiments was conducted in which a milling cutter equipped with ceramic plates with similar electrical resistivity parameters (a group of plates with R = 75 ... 100 Ω) showed the following results of the milling process stability: the actual tool path without loss of processing quality was l = 32 m, after which there was a sharp deterioration in the roughness of the surface being treated and going beyond the tolerance range of the machined element, the actual period of tool life, when it could work without loss of quality was T = 35.6 min.

The examination of the treated surface showed that before the instrument reached the maximum value of operability, the quality of the treatment was within a range sufficient for details of this type - the surface roughness was not lower than Ra = 1.6 µm, the flatness deviations did not exceed 0.03 mm for a length of 100 mm .

In the second series of experiments, one of the milling cutter plates was chosen in such a way that its efficiency and specific electrical resistance differed sharply among the representatives of the group (R = 10 Ω). The tests carried out showed that the path traveled by the tool without loss of processing quality and the actual period of durability decreased by 1.5 times.

CONCLUSION

Thus, it can be concluded that equipping the cutter with replaceable ceramic plates with values of specific electrical resistance belonging to one group allows not only stabilize the milling process, but also significantly improve the tool's efficiency by increasing the period of its durability, which will ensure the required quality of the treated surfaces blanks of machine parts. On this basis, it is recommended to equip the multiblade cutting tool with such ceramic plates, performance criteria which are in the same group.

In order to increase the implementation life of the tool equipped with cutting ceramics, to increase the processing capacity and to ensure the required quality indicators, it is desirable to mount such ceramic plates in the tool, the values of the resistivity of which approach R = 100 Ω.

REFERENCES

[1] Maksarov V., Khalimonenko A., Timofeev D. Machining quality when lathing blanks with ceramic cutting tools. Agronomy Research, Tartu: Estonian University of Life Sciences, 2014, Vol. 12, P. 269-278.

[2] Maksarov V., Khalimonenko A., Olt J. Effect of porosity on the performance of cutting ceramics / Agronomy Research, Tartu: Estonian University of Life Sciences, 2016, Vol. 14, SI 1, P. 1043-1052.

[3] Maksarov V., Khalimonenko A. Forecasting performance of ceramic cutting tool. Key Engineering Materials, Trans Tech Publications, 2017, Vol. 736, P. 86-90.

Investigation Of Wear-Resistance Of Thin-Layer Carbide-Titanium Coatings Obtained By Vacuum-Arc (PVD) Deposit

Viktor A. Krasnyy, Vyacheslav V. Maksarov*

Saint Petersburg Mining University, Russia

* vikras1955@yandex.ru

ABSTRACT

A review of studies on the use of coatings from titanium carbide has been carried out. Examples are given of the use of TiC coatings to increase the wear resistance of a number of critical parts, and also as an anticorrosion coating in various fields: aerospace, nuclear power engineering and automotive industry.   The use of titanium carbide for solving a number of tribotechnical problems is considered. The peculiarities of the process of vacuum-arc deposition of TiC coatings are studied. The influence of a change in the technological regimes of vacuum arc deposition on changes in the density of coatings, as well as in the ratio between titanium carbide and individual elements is shown. The problems of optimization of TiC coatings obtained by the PVD method for wear resistance with regard to friction at high pressures with lubrication are considered. Studies have been carried out to study the effect of technological regimes of pressure variation in the working chamber and the potential applied to the substrate with a specific wear rate during friction of two rotating cylinders with lubrication.The technological modes of PVD for the formation of optimal for wear resistance of TiC coatings.

Keywords: wear resistance, carbide-titanium coatings, vacuum arc spraying, specific wear rate.

GOALS

The aim of the work was to study the influence of vacuum-arc spraying technological modes on the wear resistance of carbide-titanium coatings by changing the pressure in the working chamber and the potential applied to the substrate, with respect to friction at high pressures with lubrication.

INTRODUCTION

Titanium compounds have found wide application in various industries. This is due to the fact that it has a number of unique properties, such as super hardness, high heat resistance, high modulus of elasticity, resistance to acids and alkalis, good electrical conductivity.

93

Coatings of titanium carbide are widely used and applied in various fields: aerospace, nuclear power, automotive. TiC is one of the most effective materials used as a wear-resistant coating. New prospects for the use of titanium carbide are discovered when using it in a nanostate (modification of alloys of different composition and purpose).

For the deposition of high-quality titanium-based coatings, the vacuum-arc deposition method (cathode-arc deposition) is a widely used method of coating thin films in a vacuum by condensing material from plasma streams generated at the cathode - a target in the cathode spot of the vacuum arc of a high-current low-voltage discharge developing exclusively in pairs of electrode material. The method is also known under the names: cathodic arc deposition (Arc-PVD), CIB-cathodic-ion bombardment method. The method of vacuum-arc spraying allows a wide variation in the structure of the emerging coatings due to a change in the pressure in the working chamber and the potential applied to the substrate. The change of technological regimes leads to changes in the density of coatings, as well as in the ratio between titanium carbide and individual elements. Obviously, such structural transformations should influence the wear resistance of the coatings obtained.

MATERIALS AND METHODS

Experimental studies were carried out on a standard friction machine. The investigated stationary sample coated with TiC-roll is made of steel 20 and had a diameter of 46 mm. The rotating roller with a diameter of 46 mm was uncoated and was made of steel 45 followed by quenching (42 ... 45 HRC). The coating was formed at two pressure levels C_6H_6 - 0.17 Pa and 0.27 Pa and four potential levels on the substrate - 50, 100, 150 and 200 V. The coating thickness was 6-10 μm. To lubricate the friction unit, an oil bath with a volume of 200 cm^3 was used, into which a rotating roller was partially immersed (2 mm). As the lubricating medium used industrial oil. Samples of 40Cr steel without quenching (28 ... 32 HRC) and quenching (42 ... 45 HRC) were also tested to assess the comparative wear resistance of coatings at high pressures.

RESULTS AND DISCUSSION

The results of comparative tests of the wear resistance of chromium steel and TiC coatings applied at pressures of C_6H_6 -0.17 Pa and 0.27 Pa at stresses in the 200 V substrate showed that the specific wear rate for coatings in 6 ... 10 times lower than for hardened steel. In addition, it can be concluded that the wear resistance of TiC coatings depends strongly on the pressure in the chamber when coating is applied. The coating applied at a higher pressure has a significantly higher wear resistance. In this connection, for subsequent tests on the influence of the potential of the electrode, the samples were made at a chamber pressure of 0.27 Pa. At a load of 500 N compared with a load of 50 N, an intensive coverage of the coating is observed, but a significant difference in the wear resistance of coatings applied at stresses on a 50-100 and 150-200 V substrate is observed. The minimum value of the specific wear rate was attained at a voltage of 150 V, and the critical pressure was noticeably higher for TiC coatings formed at substrate potentials of 150 and 200 V and reached a value of the order of 110 MPa.

CONCLUSION

As a result of the performed studies, technological regimes of VDN were chosen to form optimal TiC coatings for wear resistance, some features of wear of such coatings at large and small loads under lubrication conditions were considered. The data obtained make it possible to expand the fields of application of TiC coatings to solve complex tribological problems.

REFERENCES

[1] Shiryaeva L.S., Garbuzova A.K., Galevsky G.V. Production and application of titanium carbide (assessment, trends, forecasts). Scientific and technical lists of the St. Petersburg State Polytechnic University. 2014, № 2 (195), pp. 100-108.

[2] Burov I.V, Lisenkov A.A. Current state and prospects for the development of vacuum arc welding technology for wear-resistant coatings. Vacuum technical and technology. 2002. Vol. 12, No. 1. pp.55-60.

[3] Krasnyy V., Maksarov V., Olt J.Improving fretting resistance of heavily loaded friction machine parts using a modified polymer composition. AgronomyResearch.2016, 14(S1), pp. 1023-1033.

[4]Maksarov V.V., Krasnyy V.A. The mechanisms of friction of thin-layer nano-coatingsunder conditions of fretting.Scientific and technical lists of the St. Petersburg State Polytechnic University.226, (2015),№3, pp. 111-120.

Formation Of Hydrophobic Coating On The Modified Polymers Surface

Valeria Semenova, Anna Kochetkova*

*e-mail: semenovaleria@mail.ru

Saint-Petersburg State Institute of Technology, Russia

Saint Petersburg, Moskovskiy Av, 26

ABSTRACT

Chemical surface modification of polyethylene (PE) by phosphorus oxide (-P-O-) and phosphorus-silicon-organic (-P-O-Si-R) structures was carried out by the method of molecular layering (ML). The degree of hydration of the modified samples surface in conditions of storage in air atmosphere with a constant relative humidity was controlled by using the method of atomic force microscopy (AFM). The effect of imparted functional groups in the degree of hydrophobicity of the PE surface was assessed by using the method of measuring the contact angle. The study on the presence of physicaly sorbed water on the surface of the modified PE samples, which were storing in a desiccator with a relative humidity of 85%, was carried out by the method of IR-spectroscopy.

Keywords: polyethylene, molecular layering, atomic force microscopy.

INTRODUCTION

Low cost and availability of PE make it a perspective material for the creation of film electrets. Nowadays polymeric electrets as sources of a constant electric field are widely used in various fields of science and technology. Due to the fact that the electrets are based on pure polymeric materials, have not the best properties, it's necessary to modify the film surface by structures that are able to amplify polarization effects. It is known that the formation of nanoscale -P-O-structures on the material surface leads to appearance of long-stored electret properties in polymer [1]. However, in the process of store -P-O-groups, having a high hydrophilicity, form on the film surface the layer of sorbed water from the air. It leads to rapid degradation of electret properties and makes searching the ways of hydrophobic surface modification of polymers the actual task.

MATERIALS AND METHODS

The objects of study were samples of PE film with modified surface by -P-O- and -P-O-Si-R structures, which were storing after the synthesis in desiccators with different relative humidity (50% and 85%), to determine the influence of relative humidity on the rate of hydration of the surface of the studied samples.

Synthesis of -P-O structures on the surface of PE was carried out according to the method of molecular layering [2] at temperature of 60°C by an alternate treatment of PE by pairs of $POCl_3$ and by followed vapor-phase hydrolysis. To make the surface of the modified polymer hydrophobic, after synthesis -P-O-structures the material was treated in pairs of $Si(CH_3)_2Cl_2$ under the same conditions with followed vapor-phase hydrolysis of unreacted chlorine atoms.

The investigation of samples surface morphology was carried out using scanning probe microscope Solver P47 Pro (NT-MDT, Russia) by the method of AFM in tapping mode.

RESULTS AND DISCUSSION

According to AFM research the total overlap of the sample surface with -P-O-groups by layer of sorbed water when stored in a desiccator with a relative humidity of 50% happens after 35 days. For the sample with a double system, the process under the same storage conditions takes about 2 months. According to the data of AFM research whole flooding of sample surface with -P-O-structures in conditions of 85% relative humidity occurs after ~ 21 days of storage, while in the case of the sample with a double system, the process of full hydration of the surface takes almost 3 times longer.

Thus, it was shown that Si-R-groups block -P-O-structures and greatly slow down the sorption of water from the environment on the surface of the polymeric composite. It should help to extend the operation of the electret material.

ACKNOWLEDGEMENTS

Research was partly financed from Ministry of Education and Science of Russian Federation (project No 16.1798.2017/4.6)

REFERENCES

[1] Rychkov A.A., Rychkov D.A., Trifonov S.A. Stability of electret charge in the films and fiber composite polyethylene modified by pairs of phosphorus trichloride// Scientific journal of Herzen University. V.8, No. 38. – 2007. - P. 89 - 97.

[2] Malygin A.A. From chemical reactions at the solid surface to molecular layering nanotechnology// Bulletin of Saint Petersburg State Institute of Technology (Technical University). No.1 (27). – 2007. – P. 14 - 24.

Antifriction Properties Of Nanocomposite Materials Based On Aluminum Reinforced With Carbon Nanofibers For Application In Friction Couples Operated Under Extreme Conditions

Aleksandra N. Skvortsova, Aleksander D. Breki, Oleg V. Tolochko, Tatiana S. Koltsova, Sergey E. Aleksandrov

Peter the Great St. Petersburg Polytechnic University, Saint – Petersburg, Russia,

e-mail: annelet@yandex.ru

ABSTRACT

Research in the field of creation of materials from metals reinforced by carbon nanostructures, the last decade is growing rapidly. As a rule, for the introduction of carbon nanotubes and nanofibers (lip) in a metal matrix, in particular aluminum is its hardening, improvement of wear resistance and improved anti-friction properties. The main problem associated with the manufacture of composite materials is the agglomeration of nanotubes due to their high surface energy, which can degrade the expected properties as antifriction, and antiwear. The author of this work has been considered a method of synthesis of carbon nanofibers directly on the aluminum microparticles, allowing to obtain a uniform distribution of lip in the aluminum matrix.

The paper presents the results of a comparative laboratory study of antifriction properties of composite materials based on aluminum reinforced with carbon nanofibers, the friction at the steel 12H. The dependences of the frictional force on time for the developed composite materials and matrices. Implemented comparison of the results obtained for composite materials and pure aluminum samples.

Keywords: sliding friction, friction properties, carbon nanofibers, nanocomposite material, aluminum.

MATERIALS AND METHODS

It is widely known that the task of creation of lightweight materials with improved antifriction properties is an urgent challenge in innovative economy. In this regard research in the field of creating metal materials reinforced with carbon nanostructures has been actively developing over the last decade [1].

The goals of incorporation of carbon nanotubes and nanofibres (CNF) into metal matrix are its hardening, increasing wear resistance and improving antifriction properties. The main problem associated with the manufacturing of composite materials is the agglomeration of nanotubes due to their high surface energy. This phenomenon often leads to deterioration of expected antifriction and anti-wear properties.

The authors in [2] propose novel method of synthesizing carbon nanofibers directly on aluminum microparticles. It provides uniform distribution of CNF in the volume of aluminum matrix.

RESULTS AND DISCUSSION

The results showed that the aluminum samples obtained at different technological parameters, show different friction properties in the mode of translational sliding friction on a plane of steel 12H. Antifriction action of carbon nanofibers is shown in the moment of breakaway and during the movement of the friction surfaces relative to each other, indicating that the influence of the lip on the properties of oxide films of specific composite materials. The most stable friction coefficients were detected in the composite material ALM18+1.5% of the lip. This material was observed, the minimum value of the steady-state friction force ($\approx$6H) 25% less than ALM16, and 62.5% (2.7 times) less than the ALM102 material with greater hardness.

REFERENCES

[1] Kompozicionnye materialy na osnove alyuminiya, uprochnyonnye uglerodnymi nanovoloknami / A. N. Skvorcova, K. A. Lycheva, A. A. Voznyakovskij, T. S. Koltcova // Nauchnotekhnicheskie vedomosti SPbGPU. – 2015. – № 3(226). – S. 78–82.

[2] Sintez uglerodnyh nanovolokon na poverhnosti chastic poroshka alyuminiya / A. I. Rudskoj, O. V. Tolochko, T. S. Koltcova, A. G. Nasibulin // Metallovedenie i termicheskaya obrabotka. – 2013. – №10. – S. 53 – 57.

Smart Nanocomposite's letters

Volume 1

© Science Impact

Features of formation of zinc nanoparticles on substrates of glass and quartz

V. V. Tomaev[*1], *V. A. Polishchuk*[2]

tvaza@mail.ru

[1] St. Petersburg Mining University, St. Petersburg, Russia

[2] University of Information Technologies, Mechanics and Optics, St.Petersburg, Russia

ABSTRACT

Zinc films obtained by magnetron sputtering on glass, and quartz substrates are investigated. The morphology of the surface, the thickness of the films, and the size of the zinc nanocrystals were studied by scanning electron microscopy. It is established that the morphology of the surface of zinc films essentially depends both on the technological application conditions and on the substrate material. Zn film formes very small islands on initial stages of their growth on quartz substrate, but formes more significant islands on glass substrates. The geometric and thermodynamic parameters of the investigated films were estimated. Resistance of zinc films was investigated by impedance spectroscopy. The conclusion is drawn that the conductivity of the films is of a percolation nature. With increasing film thickness is observation the twinning of Zn crystals.

Keywords: zinc films, magnetron sputtering, glass and quartz substrates

INTRODUCTION

This paper reports some results relating to the thin Zn films formed by magnetron sputtering on glass and quartz substrates. The problems of obtaining thin layers of metals, both from a practical and theoretical point of view, have long been studied [1-5]. Mechanisms of growth of metal films were considered in detail in [6,7,8,9].

An island growth mechanism is realized when the condition [10]:

$$\sigma_s < \sigma_d + \sigma_{s-d} - const \times k_B T ln(\xi + 1), \tag{1}$$

where σ_s - free energy per unit surface of the substrate, σ_d - free energy per unit surface film, $\sigma_{(s-d)}$ - the free energy per unit surface substrate-film section.

EXPERIMENTAL SECTION

Zn films were deposited on glass, quartz and silicon substrates measuring 20×20 mm^2 by magnetron sputtering at various currents and constant pressure Ar $\sim1\times10^{-2}$ Torr. The source was a target of Zn with a diameter of 57 mm. The distance between the target and the substrates, which are fixed on a rotating holder holding at a constant speed, is ~5 cm. The deposition rate was measured. The film thickness was determined by an integrated meter, the value of which was taken as the "effective film thickness". In this case the effective thickness is equal to h_{ef} film: $h_{ef} = m/d\times S$, where m - mass of deposited metal, d - the density of the metal material, S - area occupied by the film on the measuring surface.

As can be seen from the SEM results in Figure 1 a, b and the histograms of the size distribution of islands of the film structure on different substrates differ greatly.

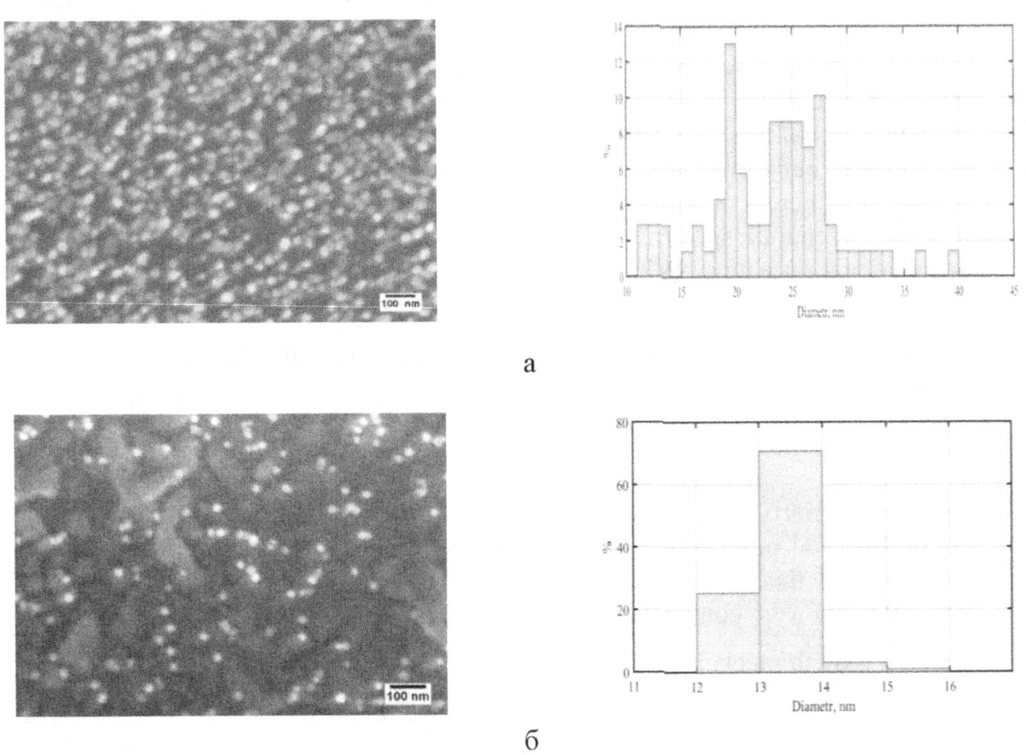

Figure 1. SEM view of the film structure on glass substrate (a) and quartz substrate (b).

The average grain size on glass is 23.1 ±5.6 nm, on quartz 12.9 ± 0.6 nm. New atoms come from the vapor of the material during the sputtering of the target. Assuming that the nanocluster is a monolayer of Zn atoms, it consists of 300-1000 atoms [11].

With a further increase in the thickness of the Zn film, the size of the islands grows on a glass substrate. The morphology of the surface does not change significantly up to a thickness of 35 ± 40 nm.

The dependence of the film conductivity on the thickness is linear.

As the film thickness increases, more zinc nanoparticles begin to touch each other and an increasing number of zinc grains begin to participate in the conductivity of the film, which ultimately leads to a decrease in film resistance.

Fig. 2 shows the dependence of the conductivity of Zn films on their thickness on quartz substrates.

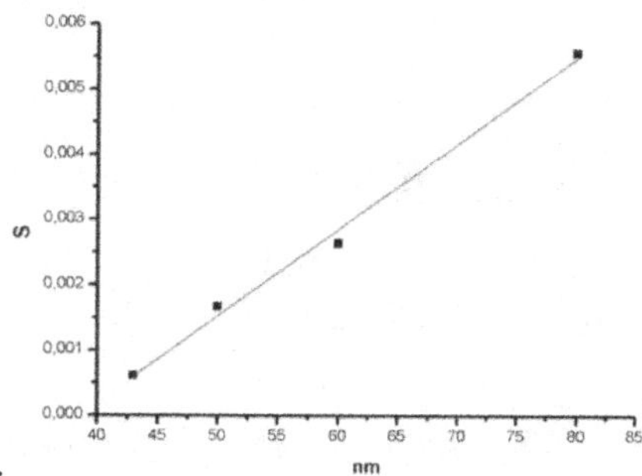

Figure 2. Dependence of the conductivity of Zn films on their thickness on quartz substrates.

For thick zinc films (h > 100 nm), the conductivity of the film will tend to conduct the bulk polycrystalline zinc.

CONCLUSION

1. It is shown that on substrates of glass and silicon during magnetron sputtering, formed island films with an effective thickness of 10÷12 nm and an average grain size of 23.1 ±5.6 nm on glass.

2. On the substrate of quartz, island films with an average size of 12.9 ± 0.6 nm islands are formed.

3.   Formed oriented hexagonal Zn single crystals with the [0001] direction parallel to the plane of the substrate.

REFERENCES

[1] Levinstein H. 1949 *J. of Appl. Phys*. V **20** April 306

[2] Konozenko I.D. 1954 *Physics-Uspekhi* V 52 N4 561

[3] Yoo M.H. 1981 *Metallurgical Transactions A* V **12A** N3 409

[4] Kukuschkin S.A.,Osipov A.V. 1998 *Physics-Uspekhi* V 168 N10 1083

[5] Ohring M. 2002 Material science of thin films Deposition and Structure. Academic Press. USA

[6] Volmer M.,Weber A. 1926 *Z. Physik. Chem.* **119** 277

[7] Stranskiy I.N,Kaischev I.K. 1939 *Physics-Uspekhi* V **21** N4 408

[8] Frank F. C. and van der Merwe J. H. 1949 One-dimensional dislocations. I. Static theory, Proc. Roy. Soc. London, Ser. A **198** 205

[9]   StranskiI.N.,Krastanow   L.1938   *Abhandlungen   der   Mathematisch-Naturwissenschaftlichen Klasse IIb. Akademie der Wissenschaften* Wien. **146** 797

[10] Venables J.A.,Spiller G.D.T., Hanbücken M. 1984 *Rep. Prog. Phys.* **47** 399

[11] Ievlev V.M., Shvedov E.V. 2006 *Physics of the Solid State* V 48 N1 133

Nanostructures with quantum well InGaAs/GaAs: experimental and theoretical analysis of tunnel contribution to emission of charge carriers

Ivanova Y.V, Zubkov V. I.*

*yana@unix-server.su

SpbSETU "LETI", Saint-Petersburg, Russian Federation

ABSTRACT

The study of heterostructures with single quantum well (SQW) $In_xGa_{1-x}As/GaAs$ was carried out by admittance methods, demonstrating two competing mechanisms of emission for carriers: thermionic and tunnel. The dependence of thermionic peaks of conductance on the reverse bias has resonance character. On the other hand, we noticed the temperature independent plateau on conductance-temperature spectra, which is always related to the tunnel nature. We guess the observed effect is the resonant tunneling through the two-barrier potential, formed at the QW borders due to the doping. Also the broadening of conductance spectra with decreasing content x in SQW is detected. The results were analyzed using the numerical self-consistent simulation.

Keywords: single quantum well, admittance spectroscopy, tunneling, thermionic emission.

INTRODUCTION

Semiconductor heterostructures containing QW are widely used in different industries, particularly in mineral and raw complex, as various light-emitting diodes, photodetectors and microwave devices. The main feature of such nanomaterials is the governing their macro properties by quantum mechanical effects. As we have shown [1], in this objects there are two competing mechanisms of charge carrier emission: thermionic emission and tunneling, which make different contribution to conductance at different temperatures.

EXPERIMENTAL SECTION

In this work, the isotype n-type heterojunctions with elastically strained single quantum well (SQW) $In_xGa_{1-x}As/GaAs$ (x = 0.065…0.29) grown by MOCVD were investigated. Thicknesses of the active layer of samples were 6.0…9.5 nm. The parameters and quality of the structures were controlled by HRXRD, local cathodoluminescence, etc. Conductance measurements were taken in temperature range from 10 to 375 K, voltage range of ±40 V, and frequency range of test signal 20 Hz – 2 MHz.

RESULTS AND DISCUSSION

The response in conductance spectra, corresponding to QW recharge, was registered only for QW with high In content in solid solution ($x > 0.25$).The temperature range for observation of response was 10-120 K. Results for structure with $x = 0.29$ (the border of pseudomorphous growth) are presented in figure 1. The temperature-conductance (G-T) spectra are experienced strong modification in dependence on the reverse bias. Peaks are expressed at reverse bias -2.3 V, which corresponds to the part of CV characteristics responsible for the beginning of intensive thermionic emission of charge carriers from quantization levels with the electric field penetrating into the QW. The observed phenomenon has resonance character and exists only in narrow voltage range – about 0.5 V. In G-T experiments the "shelf" are observed at temperatures of 20-50 K that demonstrates the existence of tunnel current, which got the maximal value at - 0.3 V, and then had sharp fall in more than 10 times of value.

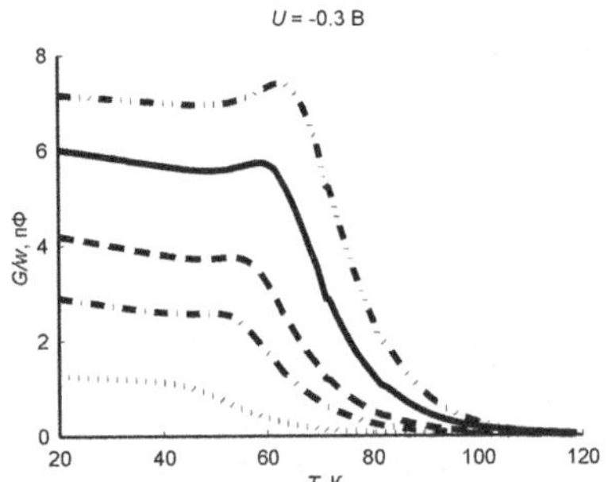

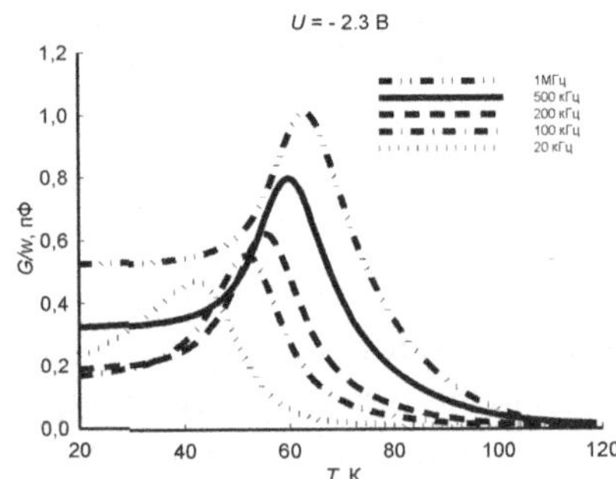

Figure 1. Admittance spectra of heterostructure with QW $In_{0.29}Ga_{0.71}As/GaAs$, obtained at U = -0.3 V (no electric field in the QW) and at U = -2.3 V (electric field approached the QW).

CONCLUSIONS

Numerical modeling by self-consistent calculation of Schrödinger and Poisson equations shown that the top part of doped heterostructures with the QW on the band diagram represents a two-barrier structure in energy region of quasi-resonances ($E > 0$) [1] with the opportunity for the resonant tunneling. For identical barriers has the maximum probability of resonance tunneling. At small reverse biases, the external electric field does not reach a space charge region yet, and the QW remains with symmetric barriers, so "shelves" on G-T spectra get the maximal values. On increasing the reverse bias the left-hand edge of space charge region begins "to rise", a symmetry of barriers is broken and the tunnel transfer probability sharply falls. In the experiments, it is expressed by sharp falling of the "shelf" at -1.3 V.

REFERENCES

Zubkov, V.I. 2007. "Characterization of $In_xGa_{1-x}As/GaAs$ quantum-well heterostructures by C–V measurements: band offsets, quantum-confinement levels, and wave functions". Semiconductors, Vol. 41, N 3: 320-326.

Properties of Er–doped (Na,K)NbO$_3$ nanofibers

Nadezhda P. Markova and Alexander M. Grishin*

Petrozavodsk State University, Petrozavodsk, 185910, Russia

khomlyk@mail.ru

ABSTRACT

In this paper, we report static electrical and optical properties of Erbium doped (Na,K)NbO$_3$ nanofibers. 100 μm long and 100-200 nm in diameter nanofibers were sintered by sol-gel calcination assisted electrospinning technique. Er doping with the concentration of 2 at.% provides readily detectable room-temperature broad-band photoluminescence centered at λ_{PL} = 0.55 and 0.98 □m being pumped, respectively, with 532 and 785 nm lasers.

Keywords: electrospinning, ferroelectric fibers, electric properties, photoluminescence.

INTRODUCTION

Ferroelectric sodium potassium niobate [(Na,K)NbO$_3$, hereinafter NKN] ceramics was patented and FDA-approved (U.S. Food and Drug Administration) as a biocompatible material for implants [1]. Thorough toxicology tests showed there was no any bacterial products (endotoxin) appear as well as viability of human monocytes was not negatively affected by the presence of NKN ceramics. Erbium doping promises new photonic applications to (Na,K)NbO$_3$ ceramics which is already abundantly supplied with piezoelectric, electrostrictive, low microwave loss voltage-tunable dielectric, and electro-optic properties.

EXPERIMENTAL SECTION

Fabrication of Er:Na$_x$K$_{1-x}$NbO$_3$ fibers was performed by the electrospinning method while formation of nanowires occurred in electrostatic field from the jet of polymer-precursor solution.

To prepare NKN precursor solution by the sol-gel method, sodium NaO$_2$C$_2$H$_3$ and potassium KO$_2$C$_2$H$_3$×3H$_2$O acetates were mixed in 2-methoxyethanol at room temperature and stirred for 1 hour to achieve a clear and transparent liquid mixture. Niobium ethoxide Nb(OCH$_2$CH$_3$)$_5$ and erbium nitrate pentahydrate Er(NO$_3$)$_3$×5H$_2$O were dissolved at room temperature in acetyl acetone, stirred for 1 hour, then added to the Na/K precursor solution, and continued stirring for 24 hours in a closed cap glass ware. Polyvinylpyrrolidone (PVP, 0.05 g/ml) was added to the precursor mixture to prepare the solution for electrospinning.

Viscous polymer-precursor jet was ejected from a syringe pump NE-300 that feeds the PVP/NKN solution at a constant rate of 0.5 ml/hour an electric field 1.8 kV/cm between metallic needle and aluminum foil collector located at 15 cm below the tip. The diameter of ejected fibers ranges from 300 to 350 nm.

After the process of electrospinning, to remove a polymer binder, collected jelly-like fibers were dried at 100°C in nitrogen atmosphere and annealed in the high-temperature furnace OTF-1200X in air at 800°C. The rate of heating/cooling was 5°C/min [2]. The diameter of fibers after annealing varies between 100-200 nm.

RESULTS AND DISCUSSION

XRD pattern of Er:NKN fibers recorded with the Siemens D-5000 diffractometer validates the predominance of perovskite $Na_{0.35}K_{0.65}NbO_3$ – like phase.

Electric properties of Er:NKN nanofibers were investigated using measurement of current-voltage *I-V* characteristics. They were traced for the planar Au/Er:NKN/Au cell with 350 μm thick Er:NKN fiber fabric placed onto glass-ceramic *Sitall* substrate as well as for the vertical sandvich Pt/Er:NKN(260μm)/Au structure. Circular 0.95 mm in diameter ohmic Au electrodes were thermally evaporated through a shadow mask on the top of nanofiber fabric specimens.

Two effects clearly demonstrate ferroelectric properties of Er:NKN fibers. The first is a clock-wise hysterethis *I-V* loop at weak electric field in the planar Au/Er:NKN/Au cell. The second was exhibited in the vertical sandvich Pt/Er:NKN/Au cell. A positive bias voltage shift occurs to nullify current at the descending branch of *I-V* curve. Also, Pt/Er:NKN/Au sandwich structure posseses a strong rectification of direct current as well as a resistance switching from low-resistance to high-resistance state with the resistance ratio of 10^2. Quasi-static *I-V* characteristics of Pt/Er:NKN/Au diode correlate with the results obtained by impedance spectroscopy carried out at frequencies up to 1 MHz.

Intensive room temperature luminescence was observed by pumping fibers with 532 and 785 nm lasers light. Radiation of 532 nm produces intensive luminescence at $\lambda_{PL} = 0.55$ μm and a weaker one within the λ_{PL} band from 0.96 to 1.11 μm. Also, it excites Raman-active internal vibrations of NbO_6 octahedra. Radiation from 785 nm laser results in the appearance of strong fluorescence lines with the wavelengths from 0.96 to 1.11 μm.

Photoexcited luminescence adds to NKN fibers great potential to serve as multimodal bioimaging, tensile and torsion sensors, energy harvesting biocompatible nanogenerators, material for bioelectronic implants, electrically polarizable scaffolds for bioengineering, repair, and regeneration of damaged tissue.

This work was supported by the RF Ministry of Education and Science, project no. 16.5857.2017/8.9 (state program) and by the Flagship University Development Program of Petrozavodsk State University (2017–2021).

REFERENCES

1) K. Nilsson, J. Lidman, K. Ljungstrom, C. Kjellman, U.S. patent 6,526,984 (4 March 2003).

2) A.M. Grishin, N.P. Markova, J. Am. Ceram. Soc. Vol. 100 (2017), p. 1051.

Investigating Influence of Technological Manufacturing Conditions on Structural Characteristics of Amorphous Silicon Layers of Heterojunction Solar Cells

Aleksei D.Maslov, Ekaterina V. Bezuglaya, Nikolay V. Vishnyakov

Ryazan State Radio Engineering University, Ryazan, Russia

ABSTRACT

The heterojunction silicon solar cells based on a-Si/c-Si was investigated by Raman spectrocopy. We analyzed Raman spectra to determine structural characteristics of amorophous film that allows to make a conclusion about degree of disordering of amoprhpous silicon layer on the basis of estimating angle of coupling deviation. It is shown that degree of disordering strongly depends on technological conditions of manufacturing amorphous layer. Correlations between such technological parameters as substrate temperature, deposition time, plasma treatment of surface and structural properties of a-Si are found.

Keywords: heterojunction silicon solar cells, Raman spectroscopy, structural characteristics, technological conditions, amorphous layer.

INTRODUCTION

Heterojunction solar cells based on a-Si/c-Si junction are promising structures for effective cheap terrestrial use. However disorder amorphous layers might restrict efficiency of such structures. Parameters of a-Si depend on technological conditions of amorphous layer manufacturing. Tthe aim of this paper is determination of correlations between technological conditions and structural characteristics of a-Si.

For investigation of structural characteristics we used widespread Raman spectroscopy method that is often used to study structural information. [1]. Compared to other methods of investigating structural properties, Raman spectroscopy is a more sensitive method to small fluctuations in structure within the first coordination sphere [2]. The obtained spectrum allows to

113

calculate root-mean-square deviation of the coupling angle from the tetrahedral position of the silicon crystal using Beeman's linear empirical relation [3]:

$$\Gamma = 15 + 6\Delta\theta, \tag{1}$$

where Γ - width of the peak, cm^{-1}; $\Delta\theta$ - root-mean-square deviation, degrees.

In addition to Beeman's empirical relation (1), the degree of ordering might be determined by estimation of TO-peak position and the ratio of TA/TO peak intensities [3].

EXPERIMENTAL SECTION

We investigated four testing a-Si:H/ c-Si solar cells manufactured with different technological conditions. Results of calculating the ordering degree of amorphous layer are shown in table 2.

Table 2 – Results of calculating structural parameters

Parameter / Sample	Angle of coupling deviation, degrees	TO-peak position, cm^{-1}	Ratio of TA/TO peak intensities
1	8,6	486	0,983
2	8,8	486	0,986
3	10,6	486	0,988
4	10,8	474	0,991

The samples №1 and №2, manufactured with different substrate temperature have different angle of coupling deviation and TO/TA peak ratio: sample №1 has lower values because this film is less disordered due to the lower substrate temperature (table 1).

Increasing of angle of coupling deviation and TO/TA peak ratio of samples №3,4, obtained with higher substrate temperature than samples №1,2, is related with influence of deposition time on formation of amorphous web.

In the case of comparison of samples №3 и №4, manufactured with substrate temperature 230 °C, sample №4 with surface bombarded by Ar^+, according to the table 2 is more disorder film due to the increase of defects states. It is confirmed by shift of TO-peak.

CONCLUSION

We found correlations between technological conditions of amorphous silicon layer and structural characteristics using Raman spectroscopy method. Substrate temperature, deposition time and plasma treatment of surface are considered. The degree of disordering of amorphous web is estimated by Beeman's empirical relation, TO-peak position and the ratio of TA/TO peak.

The experiments are made using equipment of Regional center of probe microscopy of Ryazan state radio engineering university.

REFERENCES

[1] Introductory Raman spectroscopy. John R. Ferraro, Kazuo Nakamoto and Chris W. Brown. - Elsevier, 2003. – 434 p.

[2] C. Smit, R.A.C.M.M. van Swaaij, H. Donker, A.M.H.N. Petit, W.M.M. Kessels, M.C.M. van de Sanden // Determining the material structure of microcrystalline silicon from Raman spectra. J. Appl. Phys. – 2003.Volume.94.pp.3582-3588

[3] R.L.C. Vink, G.T. Barkema, W.F. van der Weg. // Relation between Raman spectra and Structure of Amorphous Silicon. - 2000. Vol.19.pp.156-162.

The Influence Of Technology On The Properties Of Thermoeectric Materials Produced By Directional Crystallization

Sergey A. Nemov[a,b], Arseny A. Rulimov[a,], Alexandr V. Shchegolkov[c] and Anton E. Demchenko[a]*

[a]Peter the Great St. Petersburg Polytechnic University, St. Petersburg, 195251 Russia

[b]St. Petersburg Electrotechnical Univeristy, St. Petersburg, 197376 Russia

[c]Tambov State Technical University, Tambov, 392000 Russia

*rulimmmov@gmail.com

ABSTRACT

Attempts to improve the properties of $Bi_xSb_{2-x}Te_3$ solid solution produced by directional crystallization are still going on. It is argued that selenium addition increases the thermoelectric properties of $Bi_xSb_{2-x}Te_3$ samples by reducing the number of defects in Sb_2Te_3 crystal lattice. Besides the heterogeneity of the thermoelectric coefficient in $Bi_xSb_{2-x}Te_3$ samples' cross section and annealing effect on this heterogeneity are shown.

Keywords: $Bi_xSb_{2-x}Te_3$, Vertical zone melting, Heterogeneity of the thermoelectric properties, Annealing, Selenium

INTRODUCTION

Despite the long-standing reputation of low-temperature thermoelectric materials based on Sb and Bi tellurides, nowadays the theoretical base of the enterprises is being actively updated with the information of such solid solution like $Bi_xSb_{2-x}Te_3$. The main problem in the production of corresponding materials is the high sensitivity to technological parameters, which unjustified variation entails loss of both thermoelectric and mechanical properties.

MATERIALS & METHODS

$Bi_xSb_{2-x}Te_3$ solid solution is widely used in thermal converters with a working temperature range from 300 K to 600 K, wherein the contain of antimony telluride varies from 70 to 80 mol. % of the total amount of tellurides.

Directional crystallization methods have established themselves as basic for the production of thermoelectric materials including the above-mentioned composition. Growth rate anisotropy makes it possible to grow crystals with a directed structure, in which the value of the thermoelectric efficiency ZT is maximized along the growth direction [1].

DISCUSSION

During the numerous experiments we have investigated $Bi_xSb_{2-x}Te_3$ samples with the hole conductivity to detect the possible heterogeneity of thermoelectric properties. The samples had been produced by vertical zone melting. Particularly we have attempted to plot a distribution of the thermoelectric coefficient in the samples' cross section using thermal measuring probe. Typical distributions are shown in Figure 1.

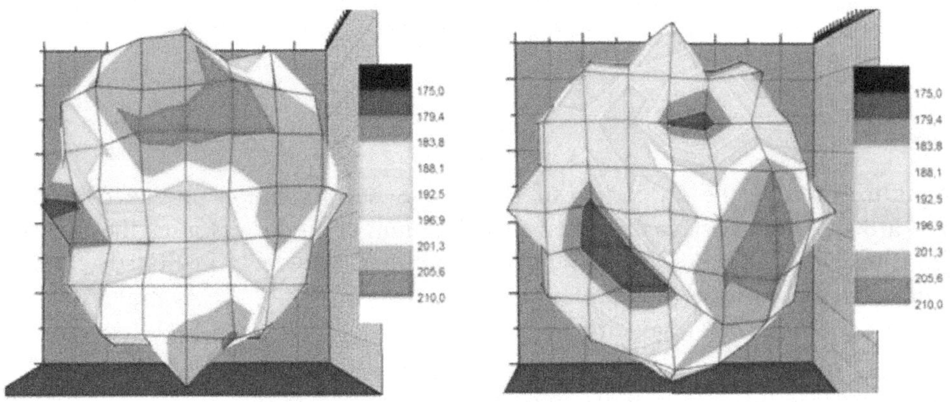

Figure 1. *The distributions of the thermoelectric coefficient [μV/K] in two $Bi_xSb_{2-x}Te_3$ samples' cross section.*

For precise adjustment of the composition and, as a consequence, useful properties of the samples and final products, it is expedient to introduce selenium into the composition, i.e. create quaternary $(Bi_xSb_{2-x})(Te_{3-y}Se_y)$ solid solution. The fact is that selenium additive smoothly reduce the hole concentration and the number of defects in Sb_2Te_3 crystal lattice, which causes an increase in the hole mobility. All of the above has a positive effect on the values of σ and α [2, 3].

CONCLUSIONS

During the research we have confirmed and described the heterogeneity of thermoelectric properties in $Bi_xSb_{2-x}Te_3$ samples produced by vertical zone melting. Consistent patterns in the distribution of the thermoelectric coefficient along samples' cross section were not found. This is explained by the presence of compositional fluctuations at the crystallization front.

Small selenium addition to the charge improves the thermoelectric properties of Bi_xSb_2-

$_x$Te$_3$ samples and the appending of carbon nanomaterials can increase the mechanical properties.

REFERENCES

1. Lavrent'ev M.G., Osvensky V.B., Pivovarov G.I., Sorokin A.I., Karataev V.V., Bublik V.T., Tabachkova N.Yu. Mechanical properties of bismuth and antimony solid solutions produced by directional crystallization and powder metallurgy. Thermoelectric materials and their applications. Reports of the interstate conference, 2015. The Ioffe Institute (St. Petersburg).

2. Ivanovs L.D., Sidorov Yu.A. Electrical properties of Se-doped Bi$_{0.5}$Sb$_{1.5}$Te$_3$ crystals. Inorganic materials, volume 34, issue 3, 1998, 222-224. Pleiades Publishing, Ltd.

3. Luk'yanova L.N., Kutasov V.A., Popov V.V., Konstantinov P.P. Galvanomagnetic and thermoelectric properties of p-Bi$_{2-x}$Sb$_x$Te$_{3-y}$Se$_y$ solid solutions at low temperatures (< 220 K). Physics of the solid state, volume 46, issue 8, 2004, 1404-1409. Pleiades Publishing, Ltd.

Experimental study of carrier distribution in CVD single-crystal diamond with delta-doped layers

Shestakova L. M., Zubkov V. I.*

*l94shes@yandex.ru

SpbSETU "LETI", Saint-Petersburg, Russian Federation

ABSTRACT

Properties of multi-delta-doped diamond are studied. The goal of this research is to measure the charge carriers' concentration profile in single crystal diamonds with four delta-doped layers. Addmitance spectroscopy was used as the principal method to measure the properties of samples at different temperatures and applied biases. The measurements showed that only the first (shallow) δ-layer was confidently observed, especially at elevated temperatures (350-450 K). Besides, the concentration profiles included a weak response from the second delta layer, significally shifted from its true location. A general conclusion is made concerning the observed charge carriers' distribution. Further study is needed to determine the responses from the deeper delta-doped layers.

Keywords: single-crystal diamond, admittance spectroscopy, delta-doping

INTRODUCTION

The search for more advanced materials and research of the methods for their application is an actual task in microwave electronics, power electronics, radiophotonics, etc. The most promising material for such applications is semiconducting diamond. However, due to the problem of charge carriers scattering on ionized impurities, the delta-doping becomes the main method for obtaining the super-high frequencies devices based on diamond.

EXPERIMENT SECTION

The paper presents the results of study a single-crystal diamond sample with four delta-doped layers grown by CVD method. The period of delta-doped layers is about 60 nm.

Measurements of current-voltage and capacitance-voltage characteristics were obtained using a computer-controlled set up of admittance spectroscopy.

RESULTS AND DISCUSSION

For measurements below 230 K, the sample behaves like a dielectric. Due to the incomplete impurity ionization in diamond, the charge under the rectifiyng barrier is collected from different depths at different temperatures; therefore at 300 K the depletion region overlapped the delta-doped layers at zero bias. As the temperature rises, the depletion region narrows, so that at $T > 400$ K it is possible in CV-measurements to register the distribution of charge carriers in the delta-layer region, Fig. 1. The position of the concentration peak (62.7 nm) corresponds fairly well to the technologically specified position, taking into account the hardware error of the CV-method. The peak amplitude reaches $5.5 \cdot 10^{20}$ cm^{-3}.

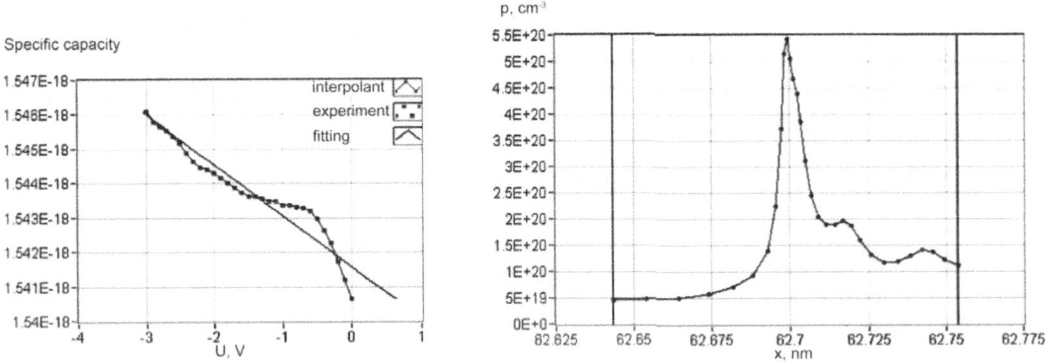

Figure 2. CV characteristics and concentration profiles in sample with delta-layers (450 K).

As it can be seen, the profiles also include the responses from the deeper delta-doped layers, but their shape and amplitude need further investigation.

CONCLUSIONS

The measurements of concentration profiles in the region of delta-doped layers in single-crystal diamond were carried out for the first time in the practice of capacitance measurements. The object of further research will be the simulation of concentration profiles by the method of self-consistent solution of the Schrödinger and Poisson equations.

REFERENCES

1.   Zubkov, V.I.; Solomnikova, A.V.; Post, J.E. et al. *Diam. Relat. Mater.* **2017.** *72.* 87–93.

2.   Zubkov, V.I.; Kucherova, O.V.; Bogdanov, S.A. et al. J. *of Appl. Phys.* **2015.** *118.* 145703.

Porous silicon-based nanostructures for solar concentrator systems

R. S. Smerdov[1]*, A. S. Mustafaev[1], Yu. M. Spivak[2], V. A. Moshnikov[2]

*rostofan@gmail.com

[1]Saint Petersburg Mining University, St. Petersburg, Russia

[2]Saint-Petersburg State Electrotechnical University, St. Petersburg, Russia

ABSTRACT

The investigation of porous silicon/silver nanostructured composite layers (por-Si/Ag) via Raman spectroscopy technique has been carried out. Interpretation of por-Si/Ag Raman spectra has been proposed. Data from layers acquired by varying anodization time showed a significant shift of a Raman peak primarily located at the frequency expected for bulk silicon towards lower energy. In addition to the single crystalline peak a number of low-frequency bands emerged in the Raman spectra of silver-functionalised porous silicon. These effects have been interpreted as due to the influence of functionalizing silver nanoparticles and synthesis parameters on the structure and properties of composite material.

Keywords: Solar concentration systems, PETE, Raman spectroscopy, porous silicon, silver nanoparticles, surface plasmon resonance, anodization

INTRODUCTION

Porous silicon (PS) and PS-based composite nanostructures are of great interest due to unique properties of such materials including photoluminescence, electroluminescence and the dependence of refraction index on porosity. It is well established that porous silicon is a promising material for numerous applications associated with micro- , nano- and optoelectronics, medicine (biosensors, targeted drug delivery), gas sensing and so on [1] by virtue of its high specific surface area and extensive capability for surface functionalization with various structures.

The use of semiconductor (GaN) cathodes in photon-enhanced thermionic emission systems has proved to be effective although only a few (less than 1%) of the incident photons exceed the 3.3 eV GaN band gap [2]. Due to the aforementioned drawback further research in the field of semiconductor materials and nanocomposites is required. Porous silicon (PS) and PS-based nanostructures in particular exhibit nearly optimal parameters in terms of the band gap (1.1 eV).

The band gap modification for PS-based composite nanostructures is possible in the wide range of 1 eV and 3 eV due to the existence of quantum confinement phenomenon and the extensive capabilities of PS surface functionalization thus providing us with an applicable material for solar concentration and thermionic energy conversion systems cathode fabrication.

Composite PS – metal materials can also act as a promising substrate for identification and characterization of various biological and chemical structures due to the surface enhanced Raman scattering effect (SERS) caused by excitation of localized surface plasmons in metal nanoparticles.

EXPERIMENTAL SECTION

PS layers were obtained by anodization of monocrystalline (111) n-type silicon wafers with resistivity of 4.5 $\Omega \cdot$cm. Anodization time varied in the range of 0.5 to 40 min. The deposition of silver was carried out from an aqueous solution of silver nitrate in a three-electrode electrochemical cell. The silver chloride electrode was used as the reference electrode. Porous silicon acted as a working electrode.

The Raman spectrum of initial porous silicon matrix is characterized by a peak located at 521 cm^{-1} that is indicative of the initial silicon sample crystalline structure.

According to Cardona [3] the size of a nanocrystalline Si cluster corresponding to a certain Raman shift could be estimated by using equation

$$d = \frac{2\pi}{10}\left(\frac{B}{\Delta\omega}\right)^{1/2},\tag{1}$$

where B=224 cm^{-1}, $\Delta\omega$ is the frequency down shift of the sample referring to the c-Si line (521 cm^{-1}). Calculated sizes of Si nanocrystals turn out to be 8 nm, 8 nm, 6.2 nm, 6.2 nm, 5.2 nm with respect to 0.5 min, 1 min, 10 min, 20 min, and 40 min anodization times.

The Raman bands at 95 cm^{-1} and 146 cm^{-1} have been attributed to the silver lattice vibrations. Duval et al. [4] states that Raman spectra of both metal nanoparticles and composite materials based on those contain a low-frequency peak the nature of which is related to the mechanical oscillations of nanoparticles under the influence of incident optical radiation. Herein by using Lamb theory for nanoparticle vibrational modes it is possible to derive an equation for the frequency of the lowest-energy spherical mode as

$$\omega_2 = 0.85\frac{v_t}{dc},\tag{2}$$

where v_t= 1600 m/s, is the transverse sound velocity, a value averaged on the different directions in the crystal [4], d - is the diameter of a particle, c - is the speed of light in vacuum, ω is the observed peak position that leads to a diameter of Ag nanoparticles deposited on PS surface of approximately 0.7 nm.

CONCLUSION

The porous silicon matrix structure before the deposition of functionalizing Ag nanoparticles closely resembles monocrystalline silicon structure. The deposition of silver nanoparticles results in the deformation of silicon lattice, accompanied by the appearance of the Raman bands being uncharacteristic for the unmodified porous silicon. Functionalization of PS matrix corresponds to the emergence of silicon nanocrystallytes, resulting in a shift of the peak intrinsic to monocrystalline silicon (521 cm^{-1}) towards lower energy. The estimation of the Si nanocrystallytes scale is performed according to Cardona theory [3] with anodization time of a silicon matrix as a parameter. The origin of low-energy Raman bands has been related to the formation of Ag nanoparticles [4] (from 0.7 nm in diameter) on the surface of a PS matrix. The increase in anodization time of silicon samples gives rise to further increase in the thickness of the porous layer and the pore diameter.

The authors are grateful to V.S. Levitsky (R&D Center TFTE, Saint Petersburg, Russia) for the Raman spectra measurements of nanostructured PS/Ag layers.

REFERENCES

Smerdov R.S., Loboda V.V., Spivak Yu.M., Moshnikov V.A. 2016 *St. Petersburg State Polytechnical University Journal Computer Science. Telecommunications and Control Systems* 3(247) 12

Schwede J.W., Bargatin I, Riley D.C., Hardin B.E., Rosenthal S.J. Photon-enhanced thermionic emission for solar concentrator systems. *Nature Materials*. 2010. N 9, p.762-767.

Cardona M. 1989 Superlattices and Microstructures 5 27

Duval E., Boukenter A., and Champagnon B. 1986 *Phys. Rev. Lett.* 56 2052.

Thermodynamic Modeling Of he Heating Of Fullerene C_{70} In n Inert Atmosphere

Nick M. Barbin,[1,2,3] Vasiliy P. Dan,[1] Dmitriy I. Terentiev,[1] Sergey G. Alekseev,[1]

[1]Ural Institute of State Fire Service of Emercom of Russia, Ekaterinburg, Russia

[2]Ural Agrarian State University, Ekaterinburg, Russia

[3]Ural Federal University, Ekaterinburg, Russia

ABSTRACT

Thermodynamic modeling of the behavior of fullerenes C_{70} by heating in argon at atmospheric pressure.

KEY WORDS: warming, carbon nanomaterials, fullerene, thermodynamic modeling.

*Corresponding author

Prof. N.M. Barbin
Prof. of Ural Institute of SFSE Russia
Ekaterinburg, Russia
E-mail: nmbarbin@mail.ru

INTRODUCTION

The study of the thermal stability of condensed fullerene structures is of interest for a range of practical applications of fullerenes, and for development of technologies for their production and purification [1]. Various sources considered temperature stability of fullerenes, however, due to discrepancies in the results a clear conclusion cannot be done.This paper studied the behavior of C70 fullerenes under heating in argon at atmospheric pressure.The research was conducted using thermodynamic modeling.

THE METHOD OF CALCULATION

Thermodynamic modeling consists of the thermodynamic analysis of the equilibrium state of the system as a whole. One of the most developed and effective programs that implement such thermodynamic calculations, is the software package TERRA [2].

Calculations of the composition of the phases and characteristics of the equilibrium are performed using reference databases [3-6]. Thermodynamic modeling was successfully used in [7, 8].

RESULTS AND DISCUSSION

Computer experiment allows to determine the phase distribution of carbon in quasi-equilibrium system C_{70}-Ar on the whole considered temperature interval. Calculation methods defined temperature interval of thermal stability of the fullerene C_{70}.

In the temperature range from 473 K to 2273 K the content of condensed C_{70} in the system C_{70}-Ar remains unchanged at 0,238 mol/kg In the temperature range from 2373 K to 3273 K the content of condensed C_{70} is reduced to 0,233 mol/kg. This is due to the beginning of the process of sublimation of the condensed component of the system and the emergence of gas C, C_2, C_3, C_4, C_5. In the interval from 3373 K to 3873 K content of condensed C_{70} decreases to 0,077 mol/kg At a temperature of 3773 K observed inflection point. Further increase of temperature resulted a linear decrease in the content of condensed C_{70}.

In the interval of temperatures from 2273 K to 3973 K the content of gaseous C and C_2 increases to 0,855 mol/kg 0,825 mol/kg At a temperature of 3973 K we observe inflection point of curves of gaseous C and C_2 from the temperature. This is due to the rapid decrease in the content of condensed C_{70}, as well as a decrease in the content of gaseous C_3, C_4, C_5 in the system C_{70}-Ar. The increase of temperature from 3973 K to 4273 K leads to a decrease in the intensity increase of the content of vapour and C_2; the maximum value is achieved at a temperature 4273 K – 2.178 mol/kg and 1,391 mol/kg, respectively.

Vapor content C_3 in the temperature range from 2273 K to 3873 K increases and reaches 2,887 mol/kg At a temperature 3873 K observed inflection point. In the temperature interval from 3873 K to 3973 K intensity of increasy of the vapor content of C_3 is reduced; the content of gaseous C_3 reaches the maximum value, 4,194 mol/kg. Increasing of the temperature to 4273 K resulted to a linear decrease in the content of gaseous C_3 to 3,650 mol/kg.

The content of C_4 and C_5 vapors in the temperature range from 2273 K to 3973 K increases and reaches 0,072 mol/kg 0,256 mol/kg at temperatures 3973 K in the graphs of dependence of the concentrations of the gaseous C_4 and C_5 is observed in the inflection point caused by a decrease in the content of condensed C_{70} and increased concentrations of gaseous C and C_2. In the temperature range from 3973 K to 4273 K content of gaseous C_4 and C_5 decreases linearly to 0,054 mol/kg 0,105 mol/kg, respectively.

CONCLUSION

As thermal stability was taken the temperature at which 10% of the total number of condensed fullerenes C_{70} has undergone dissociation.

According to the results of a computer experiment, fullerene C_{70} retains thermal stability up to a temperature 3673 K.

References

[1] Borodin V.I., Trukhacheva V.A. Thermal stability of fullerenes // Modern high technologies. – 2004. – No. 1. – P. 82-84.

[2] Vatolin N.A. Moiseev G.K., Trusov B.G. Thermodynamic modeling in high temperature systems. M.: Metallurgy, 1994. 352.

[3] Moiseev G.K., Vatolin N.A. Computer simulation of the formation of various condensed forms of carbon // Journal of physical chemistry. – 2002. – T. 76. – No. 8. – Pp. 1366-1370.

[4] Moiseev G.K., Vatolin N.A. Evaluation of the thermodynamic properties of a number of condensed carbon compounds // Journal of physical chemistry. – 2002. – T. 76. – No. 3. – S. 424-428.

[5] Moiseev G.K., Vatolin N.A. Thermodynamic properties of some gaseous fullerenes // Journal of physical chemistry. – 2002. – T. 76. – No. 2. – P. 217-220.

[6] Moiseev G.K., Vatolin N.A. An estimate of the standard enthalpy of formation (sea) of metastable condensed small coassociation of carbon and certain metals // Doklady of the Russian Academy of Sciences. – 2003. – T. 392. – No. 5. – P. 653-656.

[7] Barbin Nick M., Dan Vasiliy P., Terentiev Dmitriy I., Alekseev Sergey G. Modeling the Behavior of Carbon Nanoparticles C44 When Heated in an Argon Atmosphere. Computer Experiment. - Smart Nanocomposites 2016. Volume 7 Issue 1.

[8] Kolbin T.S., Terentiev D.I., Barbin N.M. Alexeev S.G. Behaviour of Eu, Pu, Am at burning radioactive graphite in oxygen atmosphere at different pressures // Technosphere safety. – 2013. – No. 1. S. 30-36.

Theoretical Investigation of Atomic, Electronic Structure of Subporphyrinazin and Subphthalocyanine

A.S. Kholtobina[*], E.A. Kovaleva , P.O. Krasnov

Department of Physical and Inorganic Chemistry Siberian Federal University, Krasnoyarsk, Russia

ABSTRACT

Atomic and electronic structures of subporphyrinazin and subphthalocyanine were investigated by density functional theory method using planar wave basis set and PAW approximation.

Keywords: Subporphyrinazin, subphthalocyanine, DFT, density of states.

In the past decade, SubPcs have found significant application potential in various advanced technology fields such as nonlinear optical (NLO) functional dyes, sensors, light-emitting diodes (OLEDs)[1]. Subporphyrazines (SubPzs) become an attractive class of chromophore molecules because of their intense Q band absorption and fluorescence in the visible region[2]. However, in the absence of theoretical works describing their electronic properties, it is interesting to investigate their electronic structure, particularly, densities of states, using DFT.

Quantum chemical simulation of the objects under study was performed by the DFT method using the VASP program package. All calculations given in this work were carried out for 6*6*1 k-point mesh. The densities of states (DOS, total – TDOS, partial – PDOS) were obtained by the Gaussian smearing with a peak half-width of 0.02 eV.

According to the densities of states, both subporphyrinazin and subphthalocyanine are semiconductors. Analyzing PDOS, it should be mentioned that HOMO is mostly

* Corresponding Author address
Email: akholtobina-a15@stud.sfu-kras.ru

contributed by the carbon atoms while LUMO has the supply from nitrogen, boron and hydrogen atoms in the case of subphthalocyanine.

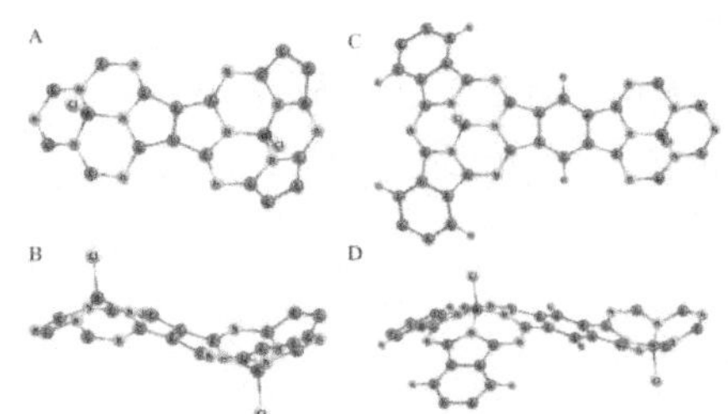

Figure 1. Geometries of Subporphirinazin (top and side views A, B) and subphthalocyanine (top and side views C, D).

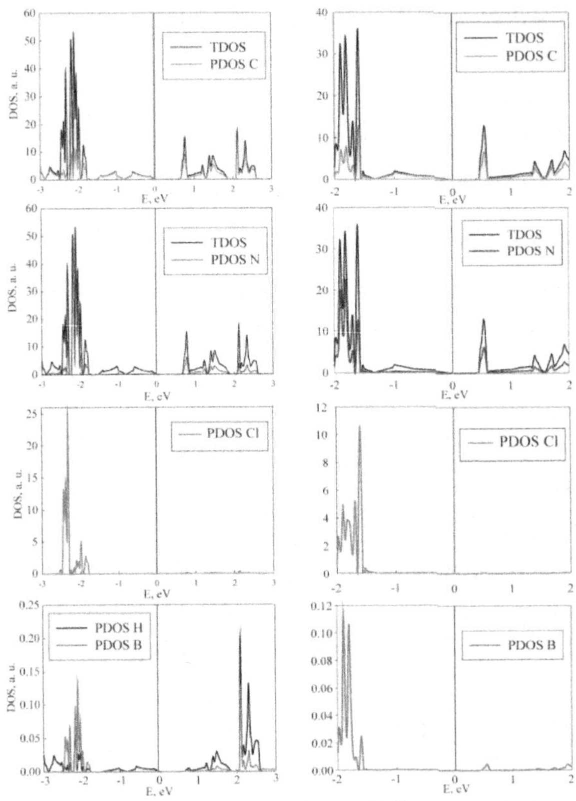

Figure 2. Total and partial densities of states for subporphirinazin (left side) and subphthalocyanine (right side).

REFERENCES

1 Wang, C., Chen, X., Qi, D., Bi, S., Jiang, J., (2017). Unraveling the formation mechanism of subphthalocyanine. Density functional theory studies. Inorganic Chemistry Communications, in press.

2 Liang, X., Shimizu S., Kobayashi N., (2014) Sizeable red-shift of absorption and fluorescence of subporphyrazine induced by peripheral push and pull substitution. Chem.Commun.,50,13781.

The Investigation Of The Linearity Of The Signals Multiplication Mixers And Amplifiers Schemes

*Yu.A. Nikitin, V.A. Yurova**

*va-yurova@mail.ru
The Bonch-Bruevich Saint-Petersburg State University of Telecommunications, Saint – Petersburg, Russia

ABSTRACT

In the paper we obtained the basic spectral and temporal characteristics of the output signal of the frequency amplifiers and mixers on the basis of which the constructed schemes of the spectrum analyzers. We present results of a comparative analysis of the linearity of the signals multiplication by using a generic approach and the same parameters of the active elements.

KEY WORDS: bipolar transistor, frequency multipliers, spectrum analyzer, spectral characteristics, temporal characteristics

INTRODUCTION

The signal mixers and amplifiers are widely used in different devices of modern electronics, in its manufacturing and the investigation of nanostructures. Therefore, the one of major task in the manufacture of electronic equipment is to provide the high accuracy of the measurement of the parameters of various materials in small quantities and in its multi-component combinations. It means that such devices as spectrum analyzers are make the strict demands on the linearity of the multiplication of the measurement signals at the circuits of amplifiers and frequency mixers, i.e. the level of nonlinearity of the active devices - semiconductor transistors, diodes etc. Generally, the different approaches, evaluation indicators and the mathematics are used in the investigations, devoted to the analysis parameters and operation modes of amplifiers and frequency converters. It's making difficulty to compare the parameters of the considering electrical circuits. So the aims of our research are obtaining spectral and temporal characteristics of the circuits of amplifiers and frequency mixers based on transistors, and a comparative analysis of the linearity of the multiplication, using a common approach and the same active elements parameters.

OBJECT OF RESEARCH

As the objects of research were chosen electric circuits based on a transistor cascade and cascade current mirror (fig. 1). In the second case it was considered two type connections of bipolar transistors in the circuit - series and parallel. These electric circuits are widely used as an individual cascades, and are the basis for designing more complex circuits of frequency mixers, which are widely used in electronics and measuring technology of the manufacturing and the laboratory investigations of nanoelectronic devices and materials of the nanotechnology [1 – 4].

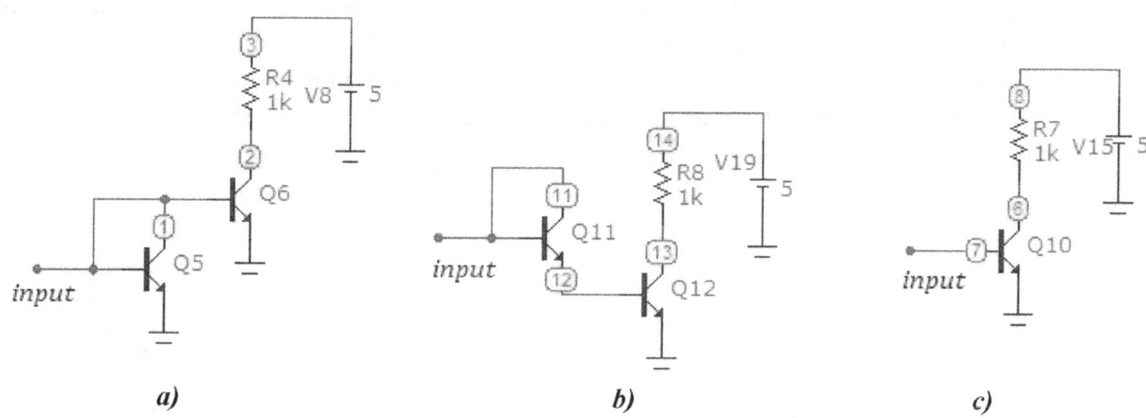

Figure 1: The diagrams of the electric circuits of cascades based on the current mirror with series (*a*) and parallel (*b*) transistors connection and the transistor cascade (*c*)

METHOD OF RESEARCH

To obtain the spectral and temporal characteristics of the output signal and to make the comparative analysis of the linearity's parameters of the cascades electric circuits was used the computer program for simulation of electrical circuits MicroCap version 9.0. Its make possible to carry out spectral research of the different structures of the electric curcuits over a wide dynamic range, applying a built-in database of the electronic components, are used in modern electronics. The criterion of linearity of the multiplies signals in mixers and amplifiers is the level of combinational components higher than the second level, because combinational component of the second level $f_{OUT} = f_{IN} \pm f_{HET}$ is the desired output signal. Here is the frequency f_{HET} is parameter of the heterodyne. If the electric circuits includes a power source (like on fig.1), so its work as an amplifier, if no, than a signal mixer.

RESULTS AND DISCUSSION

In the research we calculated parameters of the cascade electric circuits, which allow getting the specified amplitude of output signal and the transfer coefficient. Computer simulations of the frequency mixers and amplifiers schemes were made of common criteria, using the common computer simulation environment. The fig. 2 shows the obtained dependence of third level of combinational components of the output signal on the magnitude of the input signal from current source of the amplifiers based on a transistor cascade and cascade current mirror. When schemes are used as signal mixers the dependence of second level of combinational components of the output signal is investigated.

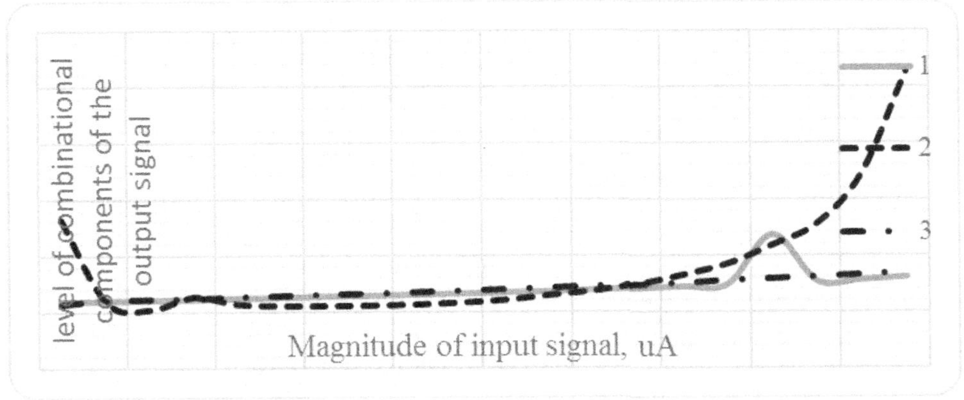

Figure 2: The dependence of third level of combinational components of the output signal of the electric circuits of cascades based on the transistor cascade (*1*) and the current mirror with parallel (*2*) and series (*3*) transistors connection.

CONCLUSION

Thus were obtained the main spectral and temporal characteristics of the output signal for electric circuits of the frequency mixers and amplifiers based on the transistor cascade and the current mirror cascade with series and parallel transistors connections. These electric circuits are widely used as the frequency amplifiers and mixers in different devices, such as spectrum analyzers, and as the basis to design more complex electrical circuits of frequency changers, such as balanced and circular mixers. The analysis of results is shown that the base cell mixer based on the current mirror provides a lower level of combinational components. It's important in develop a high-quality equipment for measuring different parameters of materials and nanoscale structures and manufacturing of modern electronic devices. Also it was found that the circuit of the current mirror with serial connection of bipolar transistors is characterized by the higher linearity, when the current generator is connected as the signal source. In the case with using voltage generator as signal source the circuit of the current mirror with parallel connection of bipolar transistors is characterized by the higher linearity.

REFERENCES

[1] Polyakov A. E., Strygin L. V. 2-Tone IP2 and IP 3 Measurement Technique. Trudy MIPT, 2012, №2. – P. 54 – 63.

[2] Gilbert B. The Micromixer: a highly linear variant of the Gilbert mixer using a bisymmetric Class-AB input stage e. IEEE Journal of Solid-State Circuits, 1997. Vol. 32, N. 9. – P. 1412-1423.

[3] Fomin N.N., Buga, N., Golovin O.V. Radio receivers: textbook for universities. - M: Techbook, 2007. – 520 p.

[4] Jones M.H. A practical introduction to electronic circuits. Third edition. Cambridge University Press, 1996. – 548 p.

Preparing cellulose-based nanocomposites and studying their electrochemical propeties

*Kirill L. Levine**

St.Petersburg Mining University, Russia, St. Petersburg

ABSTRACT

In this study, polymer coating was modified with microcrystalline cellulose (MC). The progress of water uptake inside the coating was studied by electrochemical impedance spectroscopy (EIS). Observed results show morphological changes related to swelling in ions and channels during ions propagation. Conducted study allows more justified approach in the formulation of composition of coatings for medical and technical application.

KEY WORDS: coatings, electrochemical impedance spectroscopy, microcrystalline cellulose, polymethylmethacrylate, moisture uptake, thin films

INTRODUCTION

Electrochemical impedance spectroscopy (EIS) is a non-destructive method of thin films characterization. Different porous structure can be emphasized. Impedance spectra encrypt valuable information regarding physical properties of polymer films. Decrypting this information is an important task which requires approach uniting modeling and carefully advanced physical experiment.

MC as biocompatible ecologically safe material attracts serious attention in during few recent decades [1]. Cellulose-based composites can be used in flexible polymer displays [2], biocompatible and biodegradable optical diodes [3],electro-optical properties of cellulose are investigated for application in sensors [4], such as nanocomposite with silver is studied for applications for Plasmon-base sensor [5]. In the very recent past the special attention has been drawn to biocompatible nano-cellulose for using in wet nano-cellulose membranes, artificial blood containers, materials for reconstructing bone tissue and stent coatings [6, 7, 8]. Microcrystalline cellulose is a polymer with straightened polysaccharide chain suitable for all the mentioned applications.

141

In this work processes of moisture uptake of a composite of MC with PMMA were studied in a solution close to physiological. The work was directed to establish whether biodegradable MC properties will be present ion the composite at small MC amounts, by other words, by other words, whether possible modification of properties of polymer films based on non-biopolymer by biopolymer, and how properties of non-biopolymer matrix of the composite coating will be changing.

Figure 1 shows the example EIS plot typical for studied samples. Initially samples demonstrate lower impedance, that slightly increases.

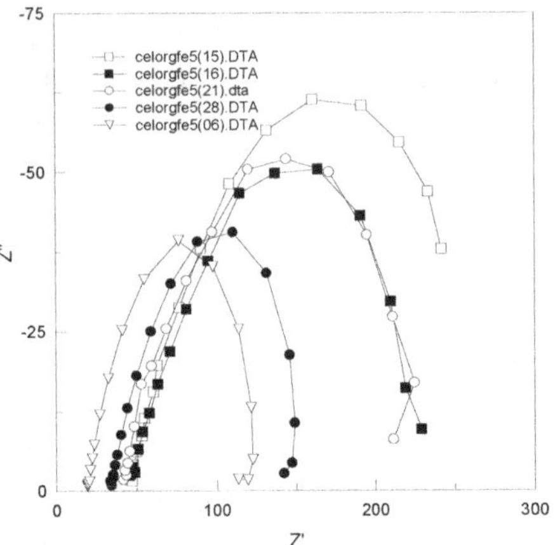
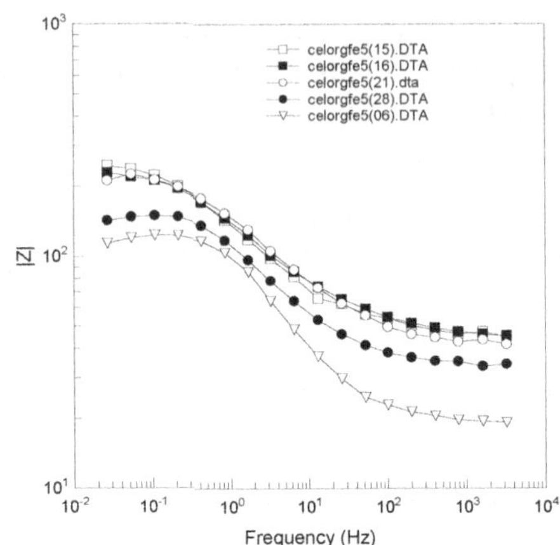

Figure 1. EIS data for PMMA with MCC film.

Similarity of plots shape is the evidence of corrosion products transport through the film rather than their storage inside film, which can be due to the presence of MCC providing micro channels for draining of those products. Those micro-channels likely to remain unchanged during corrosion that can be concluded from keeping plots shape during the entire of the experiment. Such behavior of samples containing MC allows suggesting additional mechanism of changing in the media containing ions than for classical barrier coating without MC. The properties of a barrier coting seem to be modified by the presence of MC. After the initial period of moisture uptake followed by slight impedance increasing for both barrier and modified coatings, further process of films damage for modified coatings occurs at a lowered rate and in more controlled manner, followed by keeping barrier properties longer, until decomposition.

CONCLUSION

Changes in the properties of coatings containing microcrystalline cellulose (MC) were studied by electrochemical impedance spectroscopy. Barrier properties of films change with time depending on the amount of MCC. Observed behavior can be explained by morphological changes related to swelling in voids and channels during ions diffusion. Obtained results allow more justified approach in the formulation of composition of coatings formed for medical and technical application.

REFERENCES

[1] D.A. Serqueira, A.J.M. Valente, G.R. Fihlo, H.D. Burrows, Carbohydr. Polym. 78 (2009) 402–408.

[2] Shah, R.M. Brown Jr., Appl. Microbiol. Biotechnol. 66 (2005) 352–355.

[3] C. Legnani, C. Vilani, V.L. Calil, H.S. Barud, W.G. Quirino, C.A. Achete, S.J.L. Ribeiro, M. Cremona, Thin Solid Films 517 (2008) 1016–1020.

[4] Andrey N. Aleshin, Alexander S. Berestennikov, Pavel S. Krylov, Igor P. Shcherbakov, Vasily N. Petrov, Irina N. Trapeznikova, Rustam I. Mamalimov, Albert K. Khripunov, Albina A. Tkachenko, Synthetic Metals 199 (2015) 147–151.

[5] Nahid Pourreza, Hamed Golmohammadi, Tina Naghdi, Hossein Yousefi, Biosensors and Bioelectronics, 74, (2015), 353–359.

[6] W.K. Czaja, D.J. Yuong, M. Kawecki, R.M. Brown Jr., Biomacromolecules 8 (2007) 1–12.

[7] D. Klemm, D. Schumann, U. Udhardt, S. Marsch, Progr. Polym. Sci. 26 (2001) 1561– 1603.

[8] S. Kalia, A. Dufresne, B.M. Cherian, B.S. Kaith, L. Averous, J. Njuguna, E. Nassiopoulos, Int. J. Polym. Sci. (2011) 1–35.

Made in the USA
Monee, IL
07 July 2026